MA+H POWER

How to Help Your Child
Love Math,
Even If You Don't

Revised Edition

Patricia Clark Kenschaft

Pi Press

New York

Pi Press

An imprint of Pearson Education, Inc.
1185 Avenue of the Americas, New York, New York 10036

© 2006 by Patricia Clark Kenschaft

Cartoons © 2006 by Mary Fordham

The first edition of *Math Power* was published by Addison-Wesley in 1997.

Pi Press offers discounts for bulk purchases. For more information, please contact U.S. Corporate and Government Sales, 1-800-382-3419, corpsales@pearsontechgroup.com. For sales outside the U.S., please contact International Sales at international@pearsoned.com.

Company and product names mentioned herein are the trademarks or registered trademarks of their respective owners.

Printed in the United States of America

First Printing

Library of Congress Catalog Number: 2005907609

Pi Press books are listed at www.pipress.net.

ISBN 0-13-220594-7

Pearson Education LTD.
Pearson Education Australia PTY, Limited.
Pearson Education Singapore, Pte. Ltd.
Pearson Education North Asia, Ltd.
Pearson Education Canada, Ltd.
Pearson Educatión de Mexico, S.A. de C.V.
Pearson Education—Japan
Pearson Education Malaysia, Pte. Ltd.

Table of Contents

Preface to the Revised Edition

"The more things change, the more they remain the same." This old saying reverberates as I contemplate a reprinting of *Math Power*. Several significant changes with respect to education have occurred in our culture during the past decade. However, the need for parents of young children to be involved in their children's mathematical growth is as urgent as ever. Alas, *Math Power* remains the only book by a person with a doctorate in either pure or applied mathematics for parents of children from age one through ten. As I reread what I wrote almost a decade ago, it seems as timely as ever.

Several major changes with respect to education are worth noting. One is the pervasiveness of the Internet, which was just beginning at the time of the first printing. Now you can investigate almost anything by doing a google search. However, knowing what you don't know requires some level of knowing. Moreover, nobody has "refereed" or checked what is on the web; anyone can put anything "up," no matter how ignorant or misguided. Consumers beware! Also, surfing the web does not have the continuity of reading a book, and in mathematics the connections are the core. Furthermore, a computer is not as much fun as a book to snuggle up with.

Another major change has been the increasing number of families taking their children out of public schools, sending them to charter or private schools or homeschooling them. The jury is out about charter schools, but these schools are surely exploring alternative approaches and adding zest to a sometimes too-complacent public school system. A recent study indicates that students at private schools are not doing as well on standardized tests as students with similar socioeconomic backgrounds at public schools.[1] I am sufficiently skeptical about the validity of testing not to take this conclusion too seriously, but it does help prop up the flagging belief in the quality of our democracy's public schools.

When my daughter was homeschooled for four months over twenty-five years ago, the law required a signed statement from a physician saying that homeschooling was necessary for her health. Now, parents can take their children out of school altogether with only reporting their intention, or perhaps their curriculum. The number of homeschooled American children rose rapidly in the past decade, and it now exceeds two percent of all children attending school.[2] I have added an appendix to *Math Power* especially for the large number of homeschooling parents who need to know more about the mathematics education of their children. Because I do not know nearly as much about this topic as someone whose livelihood consists of helping homeschoolers mathematically, I have been delighted to have the advice of Susan Schaeffer, who homeschooled her own three children and is now a mathematical advisor of homeschooling parents in North Carolina.

The other two major changes in our mathematical education culture are far less pleasant than the Internet and alternative forms of education. One is the split between the mathematics educators and other mathematicians. I read with sadness my statement on page 250: "On the whole these two groups work well together." It is one sentence in the previous edition of *Math Power* that I now concede to be false. In 1997, the "Math Wars" broke out. When I began hosting a weekly radio talk show, *Math Medley*, in 1998, I believed I was in between the two sides. After interviewing leaders from both sides for a few years, I became convinced that I am solidly in both camps; I want *both* accuracy in subject matter and a variety of teaching approaches. The greatly publicized misunderstanding of the "Math Wars" has been a great loss to our children. But it reflects the widespread lack of understanding of both mathematics and education that is so destructive to our country.

Finally, the use of standardized testing has increased greatly. Contrary to previous commitments to states' rights, our federal government has mandated standardized testing. I believe that those who are inflicting this anxiety on innocent children share my basic concern for education; I am not convinced by those who suspect that politicians' primary motives are to enrich their corporate test-making friends as payback for political support. However, I am amazed at politicians' naïveté in not questioning the competence of test-makers. Making up tests requires no accountability whatsoever to either a government or a professional group. How can we trust unaccountable companies to implement accountability on defenseless children?

What is the impact of standardized testing on our nation's schools? What is a good school? How do we measure that? What "should" a child know after completing a particular grade? How do we measure that? These are very difficult questions to answer. Their trivialization in much public and political discussion provides increasing evidence of the mathematical ignorance of my fellow citizens. Not every set can be "linearly ordered." In other words, you can't put them in a line, like temperatures or volumes. In particular, it is deceptive to attempt to put children's educational achievements in a line with some completely superior to others. As I contemplate my country's propensity to arrange children in order by test grades, I am ever more impressed by the wisdom of my husband's observation about quantification quoted on page 205.

I am personally most concerned about the first question: the impact of standardized testing on our nation's schools. The Educational Testing Service (ETS) is one of the most respected American producers of standardized tests, the most famous of which is the SAT. ETS recently published a booklet, *One-Third of a Nation: Rising Dropout Rates and Declining Opportunities*, that emphasizes the difficulty of measuring high school dropout rates.[3] As the stakes get higher, schools become more clever at fudging figures, and counting any large real-world quantities is not easy at best.[4] However, this publication provides convincing evidence that the dropout rate increased as standardized testing increased during the 1990s. It rose from less than a quarter to about a third of our nation's young people.[5] In ten states, the high school graduation rate declined by 8 percentage points or more.[6] The result is that although the high school graduation rate of the U.S. over-44 population is the highest in the world, the graduation rate of U.S. young people now places tenth, with fewer opportunities for dropouts than there were when today's middle-aged Americans were young.[7]

A related alarming statistic is from a study by the Texas Department of Criminal Justice in 1998: Two thirds of the inmates in the Texas prison system are high school dropouts. The cost of having fewer high school graduates is high in crass taxpayer money, but the social cost is much higher. The costs are not spread equitably across our citizenry. Nearly half of our nation's African American students and nearly 40 percent of Latino students attend high schools from which a majority of students never graduate, whereas only 11 percent of white students attend these schools.[8]

Ironically, the national legislation called "No Child Left Behind" is causing more children than ever to be left behind. Having elementary school children repeat grades is simply not done in Japan,[9] where test scores on international mathematics tests are far above ours. In Japan, it is considered cruel to remove a child from his or her friends. Thus the pressures are strong on Japanese children to help every classmate learn so there are no "drags" on the class. American research indicates that half the children who repeat a grade do no better the second time, and a quarter actually do worse.[10] I believe our national attention should be more focused on including all young people in our culture and economy, and less on marginal improvement of the quality of our schools, many of which are already excellent.

Education should involve far more than any test can possibly measure. Preparing children for tests diverts schools from the core ideas of mathematics and other disciplines, does not encourage students to look at the diverse, non-standard approaches so needed for human and national survival, and teaches youngsters that learning is only to prepare for tests instead of being intrinsic to a good life. Who needs to take a multiple-choice test to be a good employee, citizen, or parent? We are squandering our children's precious lives by teaching them a useless skill! My students at Montclair State University (which is sufficiently selective to accept only one out of every five applicants) tell me that studying for tests has taught them not to think of even *trying* to retain what they were taught after the test. Test preparation becomes the end of education, and youngsters don't think about the importance of remembering later what they learned.

Because tests that have high stakes for schools and principals are typically administered in tenth grade, students are retained in ninth grade in ever-greater numbers. Thirty-eight percent of the ninth graders in Texas public schools in 1999–2000 left school prior to graduation,[11] significantly higher than the 33 percent in 1985–1986[12] or the nationwide average in 2000. And this is the state from which the national secretary of education was recruited to craft the No Child Left Behind Act! Since then, there have been major exposés of serious cheating in the Houston public schools. Reports of cheating elsewhere proliferate. One researcher writes, "Just as high financial stakes create incentives for corporate leaders to fudge data, high stakes associated with school accountability can encourage educators to cheat on tests or otherwise game the system."[13]

Perhaps worst of all, widespread testing is driving some of our best teachers out of the public schools. They aspire to educate, but they become so frustrated by the extent to which their jobs have been reduced to rote preparation for tests that they leave. Having only rule-followers in our public schools is not good for our children or our country.

I do not know of any teacher, imaginative or rule-follower, who welcomes high-stakes standardized tests, although some are more resigned than others. Why politicians think they know better than teachers how to improve the education of children escapes me. Do they think they know more about health care than physicians? In 2002, educators in the state of Washington filed an initiative requiring any candidate running for any local or statewide office to take the same high-stakes test required of all tenth grade students and to post their scores publicly. They did not have to "pass." The goal was to help public officials understand better the process of taking the tests on which the stakes are so high for students and schools.[14] The legislation did not pass.

I do not regret any sentence in Part IV of *Math Power*. That section about the dangers of tests and grades is more true than ever.

Yet I remain optimistic about the future of American education. There are many excellent teachers in our public schools, even if some don't communicate a love of mathematics. They can teach other subjects well, and they could learn mathematics quickly if our society provided appropriate opportunities.

Our children are as smart as ever. I have a twenty-one-month-old friend who went to visit a home where one dog had lived. Since her last visit, the family had acquired another. Lila surveyed the situation and said, "Two doggies!" Lila was having fun using mathematics before the age of two. If other parents join the movement to help their children and their children's teachers, we might have a country of math-joyful citizens surprisingly soon.

Preface to the First Edition

People were meant to enjoy mathematics. This book describes an approach to math that gives both insight and delight. Through games, questions, and conversations, you too can build your own understanding of math while laying the groundwork for lifelong mathematical pleasure and satisfaction in your children.

Over the past decade, my work with children in the early grades has convinced me that *every* child can share my joy in mathematics if taught appropriately—and that parents play an indispensable role in fostering their children's innate joy and competence. Cultivating a child mathematically reminds me of cultivating a garden. Each child, like each plant, is individual, and needs to be enticed and disciplined in a special way that only a loving caretaker can see. On the other hand, young humans, like seedlings, have many common needs and growth patterns. I wrote this book to share the rhythms, tools, and seeds of ideas that I have found useful in cultivating children mathematically.

Most important is understanding the basic growth rhythms that direct your child's creative ideas in satisfying directions. Therefore, the first five chapters explore what math is, what real math ability is, and how math blooms. Too much that has nothing to do with real math masquerades as math in U.S. schools, turning children away from the subject that has brought many of us so much excitement. You want to nurture your child's innate ability and avoid using highly touted poisons that may accelerate growth temporarily but will stunt long-term health and achievement.

Which way to go "depends a great deal on where you want to get to," as the Cheshire cat observed in *Alice in Wonderland*, written in 1863 by a mathematician. I write for parents who want to help their children become mathematically happy and successful, who are willing to learn as much math themselves as they conveniently can, and who hope to share joys with their children in as many ways as they can squeeze into their busy schedules. If this is "where you want to get to," Part I and the last chapter describe the mathematical groundwork—how to build the healthy garden soil in which your child will thrive. It may be the first such explanation by a mathematician for parents, and surely is the first in recent decades in this country. Until adults understand "which way to go," children won't achieve or enjoy math as they might.

Parts II and III give detailed, hands-on suggestions about how to inspire your child mathematically, addressing the preschools and elementary school years, respectively. Some of the ideas take time, but most can be implemented during other parenting activities. They are the heart of the book, but they need the spirit and framework offered in Part I to be used effectively.

Part IV indicates why so many of our fellow citizens have missed both the fun and the power of mathematics. Customs set long ago and too rarely questioned are hard to change, but parents, teachers, and others who want to prepare our youngsters for a world based on technology must grapple with issues beyond their own homes and classrooms. These five chapters help readers understand and confront problems not of their own making.

Such problems need not continue if we all, as a united culture, address and change our societal habits. Although bickering makes headlines, math reformers of all stripes have a great deal in common. Details about dedication, enthusiasm, practice, vision, honesty, kindness, thrift, individuality, and family may be debated, but almost everyone admits they are important. The last part of the book suggests ways that groups of people, both locally and at state and national levels, can collaborate to create an intellectual garden that nourishes all our children.

Throughout the book, material not needed for continuity is enclosed in boxes. Some boxes indicate detours around mathematical concepts that may be intimidating. Others contain enrichment. The indented passages are illustrative anecdotes that may be essential for understanding the ideas and suggestions. The book's chapters can be read in various orders, since the editors and I have taken care to refer readers to other parts of the book as needed. However, please read the first section (Chapters 1–5) and the last chapter carefully at some time, since knowing where you want to go is essential for choosing some right path, and this book is based on the belief that all parents are capable of finding that special path for their own children. There is no magic formula, but wonder, love, joy, and understanding *are* available to us all.

Writing, like math, continually prods one to say, "Why do I think that?" and "Is there another way?" There always is another way! Yet the time has come in the writing of this book to say "enough," and to hope that I will be forgiven for the better ways that might have been found if I were more perceptive, had worked harder, or had consulted more people. I hope and pray that my enthusiasm may be contagious and that other families, each in its own precious unique way, will find delight in mathematical exploration, each repeatedly asking, "Is there another way?"

Acknowledgments

Because this book is a fulfillment of four decades of dreaming, it owes debts to hundreds of people, only a few of whom I can name here. Foremost, of course, are my parents and offspring, with whom I shared the adventure of math power at home: John Randolph Clark, the late Bertha Francis Clark, and Lori and Edward Kenschaft. Almost as obvious are those who read all or part of the manuscript and made suggestions: Diana Autin, Camille Barowski, Karen Bernard, Joan Buchese, Fred Chichester, Rudy Clark, Sue Geller, Suzanne Granstrand, Pat Hess, Lori Kenschaft, Norma Kimzey, Lee Lorch, Sara Mastalone, Angela McBride, Nancy Mehegan, Ken Millett, Joe Morton, Henry Pollak, Kay Pruett, Ruth Rosenblatt, Edna Smith, Max Sobel, Lynn Steen, Rosemary Steinbaum, and Susan Stillman.

The tirade in Part IV is not out of my direct experience, but from listening to many others; I consider my own happy life the result of good parenting and teaching. The best of my own fine elementary school teachers, Irene Weyer, never let her "little people" guess she wasn't in love with math until one became a math professor. She and I have exchanged Christmas cards for 50 years! Maxine Hoffer, my creative writing teacher at Nutley (N.J.) High School, graded papers *every day* for each student, thereby stimulating me to combine writing with math. Alice Brodhead's education courses at Swarthmore College not only excited us about education, but also demonstrated how fabulous education courses can be.

All of my high school, college, and graduate school math teachers were men, but I can't remember any of them hinting that my sex might be a liability. Many were superb teachers; the worst welcomed my assistance. The late Edward Assmus, Sr., taught me in high school that math was both harder and easier than I'd thought. David Rosen, my advisor and math professor at Swarthmore College, assured me that an early marriage need not thwart an ambitious career, and always encouraged me to be "innovative." The late Emil Grosswald invited me to get a doctorate "as soon as possible," and arranged for a teaching fellowship at the University of Pennsylvania. Edward Effros, my Ph.D. advisor, suggested when I finished my dissertation in C*-algebras that I might want to do something unusual with my degree; even this year he encourages my rebellious career path.

I continue to be lucky. Montclair State University is a community of people who really care about learning and sharing. I may have the best job in the world. In particular, Kenneth Wolff, my department chair for over fifteen years, tolerates folks who see the world as it might be, and manages to channel their dreams in ways that do minimal damage to the world as it is. Louis Giglio, a visionary high school teacher, expressed his midlife crisis in an urge to reform nearby elementary math education, and suggested we collaborate in starting PRIMES, the Project for Resourceful Instruction of Mathematics in the Elementary School; as Executive Director he prevailed over financial and other exasperating details. Assistant Director Marilyn Hughes had a knack for solving problems before I knew they existed, and the other PRIMES leaders found marvelous complementary niches. Hundreds of elementary school professionals seized PRIMES' enthusiasm and spread it in their schools. For example, Valerie Miller and Susan McBride, who I met when they were third grade teachers in Newark's Broadway Elementary School, witnessed my first inept attempt to reach elementary school teachers and helped me explore better ways, while learning to perform miracles in their own classrooms. My MAA, UNA, AAUP, environmental, and religious activities have introduced me to thousands of inspiring people who radiate hope for the future of humankind.

Recently, my good luck in having excellent mentors continued in my agent, John Wright, and my editors, Elizabeth Maguire and Julie Stillman. Without them and the fine staff at Addison-Wesley, this book would have remained only a dream. My neighbors have been wonderfully supportive; two of them posed with my garden tulips and me on the park bench at the end of our block for the back-cover photo. The many folks I telephoned sight unseen with questions about the book's content were invariably cooperative. I hereby thank everyone named in the book.

Most important has been my husband, Frederick D. Chichester, who not only does all the laundry and shopping (and patiently listened to the entire manuscript), but is my best advertiser and a constant source of inspiring ideas. Best of all, he's fun to live with. And yes, we still love to do math together.

EMPOWERING YOU

Math Power:
Who, What, Why, and How

1

Average students in other countries often learn as much mathematics as the best students learn in the United States. Data from the Second International Mathematics Study (1982) show that the performance of the top 5 percent of U.S. students is matched by the top 50 percent of students in Japan. Our very best students—the top 1 percent—scored lowest of the top 1 percent in all participating countries. All U.S. students—whether below, at, or above average—can and must learn more mathematics.

—*Everybody Counts* [1]

Mathematical joy! What a gift for your child!

Math power is the ability to use *and enjoy* mathematics. Math power gives a feeling of control, both over ourselves and our environment. It is a valid feeling. If we understand math well enough to use it spontaneously, we do have greater control both over our inner life and over society's decisions.

Children are born with enthusiasm for math. Think of how much your preschool child enjoys counting! It enlivens humdrum activities like stair climbing, toothbrushing, and toy-sharing. Or watch your child arrange toys in a pattern; that too is exploring math.

Every parent and caretaker can help preserve young children's innate enthusiasm. You don't have to be trained, or certified, or rich, despite rumors to the contrary. You *can* do it, whether you are an upwardly mobile professional, a struggling single parent, a home-schooler, an environmentalist retreating to the simple life, or Aunt Gertrude, whose family just dumped the kids on your doorstep.

A supportive family is the single most important factor in the intellectual success of their offspring, even if the parents have only a fourth grade education and regardless of whether they remained married throughout the youngster's childhood. Mathematical competence is helped by parental enthusiasm for learning, easily available books of all types (not just math), and habits of family conversation. Most of these require remarkably little money, especially in a community with a public library.

Although I do not believe support can be measured, I do believe it is worth writing a book about how to provide it. The most important things in life can't be measured. Mathematics has its limits.

Studies repeatedly show that the greatest *measurable* predictor of a child's academic success is the parents' socioeconomic status. Without doubting the conclusion of these expensive studies, we can be skeptical about their meaning and application. We all know of high socioeconomic parents with a shiftless offspring. Conversely, I know of a minority single mother waitress whose female child grew up to be a college mathematics professor.[2] Measurable correlations cannot be legitimately used to predict individuals' fates.

Black mathematicians of New Jersey: In the mid-1980s I surveyed African Americans in New Jersey with degrees in math. There were 75 responses. All had professional careers.[3]

- ▶ Only 4 had two parents who were college graduates.
- ▶ The majority, 44, did not have two parents who finished high school.
- ▶ Almost a third, 23, had no parent who had finished high school.
- ▶ Almost a quarter, 18, had no parent who had begun high school.[4]

Obviously, formal education is not necessary for raising a professional. There is much more involved than formal education, social status, and economics. Most of these parents were near the bottom of the socioeconomic ladder. Indeed the most mathematically successful parent that I have ever heard of was the daughter of two slaves.

A Slave's Daughter: In 1912 Susie Johnson McAfee took the examination for Texas teachers' certification. She passed every part except that her spelling paper was mysteriously lost. Her father was told that if he paid $50, the paper could be found. He said he was outraged and refused to pay, so his daughter couldn't become a teacher.

She married a carpenter-farmer and had nine children. "She taught *us*," said one son, Dr. Walter McAfee. Five of her children received degrees in mathematics itself. Another successfully completed two years as a math major and then became an electrician, later obtaining a degree in his chosen math-related field. Two others earned degrees in chemistry.

That makes eight. The youngest rebelled and earned his undergraduate degree in physical education. In middle age, however, he took more courses and became a mathematics teacher.[5]

Susie Johnson McAfee was not high in socioeconomic class, but she is a superb example for the skeptical or discouraged. Admittedly, she may have had some unusual intellectual resources, but she obviously also had serious obstacles.

Our country seems mired in destructive teaching techniques. As the opening quote indicates, a top U.S. high school student at the 95th percentile has learned mathematics comparable to a Japanese youngster in dead center. One reason is our structural bias against real mathematics. Twenty-four states have no math requirements whatsoever for prospective elementary school teachers, and the others require passing only a minimal test or course.[6] Yet all non-handicapped children can learn, regardless of sex or ethnicity. Until states require elementary school teachers to learn more math, parents are their children's major hope.

In other times and places it was assumed that people would learn if taught. Plato's *Meno*, written over 2,300 years ago, relates how a slave boy learned mathematics merely by being asked questions. Plato argued that we all knew math from our previous life and that it need only be recalled. His theology is debatable, but his conclusion of human teachability has been verified again and again.[7]

Mathematical power is in jeopardy in our culture because the popular assumption is "that even with hard work some children are doomed to low levels of achievement."[8] Harold Stevenson and James Stigler, who compared classes in Minneapolis and Chicago to classes in Japan and China, claim that, despite U.S. rhetoric about equality, our educational system assumes that math power depends more on innate ability than work. They argue that parent involvement through high school is critical and that although school reform is necessary, it will not succeed unless parents "become more involved in their children's education and develop appropriate expectations."[9]

Fortunately, parents can help their children without a great deal of formal background themselves. It takes some work, of course, but usually it will be fun, and parenting is not always easy. Keeping your child from mathematical misery may prevent serious difficulty in the long run.

Our children need courage to tackle complex ideas. If you follow the guidelines of this book, I can't promise that your child will be outstanding, but I am quite sure that he or she will be comfortable with math. Unless you insist upon A's (or your child falls into one of the scholastic traps that I will describe), you and your youngster will be satisfied with her or his grades. If you welcome more math into your home life, your entire family will be glad that you did.

What Is Mathematics?

Mathematics is the study of patterns and the use of patterns to solve problems. Mathematics is a language. Mathematics is a spirit. It is shared most effectively as a journey with love, joy, and wonder.

Math power evolves from a combination of knowledge, ability, and attitudes. These are closely related. Without knowledge, we can't develop ability. Without ability, we can't use knowledge. Without a hopeful attitude we won't acquire either math knowledge or ability.

Computation (routine calculation) is to mathematics as spelling is to literature. It has value in itself, but it is no substitute for the real thing. Just as good literature entices young children to read, real mathematics should be provided at the earliest stages of mathematical involvement. If it is not, children get a mistaken belief about what math is. Too often this belief haunts them for a lifetime.

As real mathematics struggles for survival in our culture, it becomes increasingly urgent that parents share real mathematics with their own children. Patterns are everywhere. You can enjoy showing your child relationships between patterns in one place and those in another. If you are alert, as you will be after reading this book, you can find many ways to make mathematics come alive in your child's life—and your own.

What are patterns? Patterns involve relationships, resemblances, and rearrangements—noticing what is the same and what is different in various contexts. Patterns occur in geometry, in music, and in human behavior. Numbers may measure these patterns, but math includes much more than just numbers.

We must use patterns to cope with the overwhelming details in our lives. Nobody can process all the data we experience daily. Math power helps us resist being crushed under the onslaught. It is frightening how often mathematical nonsense creeps into public discussion.

- ▶ A politician attacks his opponent for raising taxes *more times* although the *number of times* that a person supported a tax increase does not affect our pocketbook. The total *size* of the increases, along with how the taxes and the services they buy are distributed among the taxpayers, determine how they affect our family spending power.

- ▶ An insurance company refuses to renew a customer because of the number of *times* the customer has collected from the company, even though each collection is tiny. Just one collection from a typical customer costs the company far more.

- ▶ Some states have passed laws requiring additives in fuels to decrease the *percentage* of pollutants in the emissions. However, since the additives also lower the fuel mileage, the *amount* of pollutants remains constant. Consumer prices rise and the emissions with additives appear to be more harmful medically.

- ▶ Some campuses and municipalities have needlessly frustrating traffic patterns. Some applied mathematical analysis could enable vehicles to travel much more efficiently.

Mathematics is one way of understanding the world. It is not the only way, and I wouldn't claim it is the best way. But without it, our perceptions are incomplete.

Why Is Mathematics Important?

Why focus specific attention on children's mathematical growth? There are personal, interpersonal, career, and civic reasons.

PERSONAL

Mathematics was meant to bring understanding and joy. Most mathematicians are enviable people. Their ability to "center" affects the rest of their lives. Only one of the 75 black mathematicians in the New Jersey study wasn't enthusiastic about his or her career. (She made over $60,000 in the mid-1980s; one suspects she could have made changes if she was really unhappy.)

Intellectual growth gives inner freedom. If I really believe that nobody can fetter my mind, then there is no limit on what I can think or do. Some people can cultivate such freedom in amazing circumstances. Harriet Tubman and Dietrich Bonhoeffer are two memorable examples from the horrors of U.S. slavery and German Nazism, respectively. We can

wonder how they managed to maintain their independent perspective and courage under such circumstances. But most of us are not so inwardly strong, especially as children.

The usefulness of mathematics in everyday life is undeniable, but overemphasized. "What if literacy were taught *only* by means of parking tickets, job applications, tax forms, and other material that people will *need* to read? That would be an accurate analogy to much of the traditional curriculum in mathematics." Thus mathematician Dr. Neal Koblitz and computer scientist Dr. Michael Fellows argue for including entertaining, enticing math topics in the primary grades.[10] People who don't do math quickly too often fall prey to the vultures of society, but there are happier reasons for learning math.

In this culture, math comfort strongly affects an individual's identity. Mathematical confidence generates self-esteem useful in many facets of life. Conversely, feeling mathematically crushed curtails both private and public options. It goes far beyond competence while cooking, driving, making things, and managing finances. The sense of defeat felt by people uncomfortable with mathematics is hard to overestimate.

Even if you have been a victim of an anti-mathematics system, your child need not have a similar fate. Parents and teachers can unfetter talent in their little ones that was stifled in themselves. As they do so, the adults also grow, but usually not as fast or as far as their children.

INTERPERSONAL

Mathematics, like most interesting activities, builds bridges between people. The international mathematics community extends remarkable trust and caring among its own. For years my best insight into what was happening in the Soviet Union was via the *Notices* of the American Mathematical Society. When Canadian mathematician Lee Lorch arrived for his first scientific visit to the Soviet Union in 1966, he was warned by a Canadian diplomat in Moscow that he would have a very lonely four months because Soviet mathematicians would be afraid to have Western ties. Dr. Lorch went directly from the embassy to his hotel room, where there were already several invitations from Soviet mathematicians to their homes. In each Soviet city it was the same, and the political views of his hosts varied widely. Mathematics was enough to begin deep friendships.

People connect through specific mathematical enthusiasms. For example, people who enjoy chess, bridge, fantasy games, and computer hacking have created their own subcultures. More broadly, the ability to perceive and organize patterns enhances the speaking and writing skills that are so valuable for connecting with others.

Parents and children can develop a wonderful bond through math. You can establish collaborative problem-solving habits. When your child is a successful adult, you will feel a

International Trust

I will always remember one lunch at the 1986 International Congress of Mathematicians.

Five mathematicians from five different cultures were sitting around my table. None of us had ever met before. I was the only woman. The conversation was remarkably open, but even so, I was surprised when the Middle Eastern Muslim turned to the Indian Hindu and inquired about his vision of God. After explaining his pantheistic view of God pervading all objects, the Indian turned to the East German and asked, "Can you believe there is a God separated from our immanent world?"

"No," said the East German, "but I give ten percent of my income to the church." We all gasped.

"Why?!!!" we said in chorus.

"Two reasons." He grinned at our amazement. "See that building?" He gestured out the window. "Buildings in exactly that style are being built all over the world. I want to preserve the beautiful churches that were built in a time that cherished individuality."

We nodded. There was a long pause. The second reason? He looked at each of us searchingly. We were fellow mathematicians, and he decided to trust us.

"In my country the churches are the only place where groups of people can talk without being spied on." It seemed a long time while we all waited. We knew he had something more important to say.

"You people cannot imagine how badly people in my country want to leave." There was another pause.

"Why don't you stay here now?"

"My family. They are all there. I don't know what would happen to them if I don't return."

The trust in that diverse but unified group is rarely matched.

sense of satisfaction as you reflect upon your mathematical sharing long ago when she or he was a child.

CAREERS

As our world becomes ever more complex, career opportunities increasingly depend upon mathematical enthusiasm. Surveys consistently indicate that four of the most promising careers are mathematical—actuary (and other types of financial analysts), accountant, computer specialist, and statistician.[11] Other mathematical careers include telephone networks, drug research, and scheduling. Environmental jobs include setting standards, helping industry comply with them, helping governments enforce them, and testifying during litigation. New York City saved a great deal of money by hiring mathematicians to plan where to place its fire stations and how different teams should respond to fires in specified locations.

Careers in science, engineering, and finance do not require a math major but expect comfort and competence in mathematics. Majors not requiring math tend to lead to low-paying jobs. Secure jobs with decent pay usually require mathematical competence.

Even traditionally female careers require math for top positions. It is not just that credentials for leadership in nursing, social work, and elementary education require mathematics and statistics courses. Also, the actual job requirements of analyzing information, allocating resources, planning schedules, and controlling budgets draw upon the information and skills learned in these courses. Furthermore, leaders in any field need to contemplate overview patterns and to solve problems analytically, abilities developed by "real" mathematics.

CITIZENSHIP

Mathematics helps unmask false prophets who continually try to deceive the math-anxious. To make intelligent communal decisions, people need to know math. For over two thousand years all educated people studied mathematics because it empowered the mind and was considered essential for wise citizenship.

Now that we all may be voting citizens, we all need mathematics. Just understanding a newspaper requires math knowledge.

The success of a modern democracy depends upon the ability of many citizens to analyze complex social and economic patterns. Such analysis requires not just numbers but sophisticated mathematics.

How Do People Learn Mathematics?

This, of course, is the theme of the entire book, but we can give a brief glimpse here. Nurturing mathematical power can be one of the most rewarding ways to ease a child's entry into this troubled world. Children need parents and teachers to preserve their inborn love of learning, to empower them to learn facts, and to allow them to continue thinking for themselves.

Skills are the tools with which we do mathematics or read literature. However, if we are chained to our tools, we can't put one down to pick up another. Our creativity is hampered, perhaps mortally.

Good mathematics, like good literature, should be available to beginners because doing real math is essential for developing real math power. Teachers who haven't been taught

'illions

thousand, million, billion, and trillion

Each is a thousand times larger than the previous number, but how can we *understand* such overwhelming size?

How long is a thousand seconds? (You and your child could investigate correct ways to find the answer.) 16 and ⅔ minutes.

How long is a million seconds? About 11 days.

How long is a billion seconds? About 32 years.

A trillion seconds ago Neanderthal folks were walking the earth.[12]

How many thousands of dollars are in your household's annual budget?

How many millions of dollars are in your municipality's budget?

How many billions of dollars are in your state's budget?

As this is written, the annual U.S. budget is about 1.7 trillion dollars—and its debt is about four trillion dollars.

To control all these costs, more people must try to comprehend these numbers and must contemplate their comparative values. Only then can we get a sense of proportion about how we are spending our resources.

Corporate Decision-Making That Affects Individuals

▶ Medical tests, such as those for cancer and AIDS, always have a percentage error; comprehending the implications can be a matter of life and death.

▶ Both the safety and effectiveness of new medicines are measured in statistical ways involving error, and patients have a right and desire to know the risks.

▶ Environmental decisions affecting the health and safety of us all involve profound mathematical considerations, including extrapolations from limited observations and evaluations of risks.

real mathematics may resort to mind-deadening, repetitive drill on math *skills*. It is as if teachers who had never heard a melody devoted music classes to drill on reading musical notes. The math community refers to this boring, deadening drill as "Drill and Kill." It is very different from mind-building drill. Subjected to "Drill and Kill," students lose both the motivation and the "space" to grow and think.

Drill and Kill has spread like an intellectual disease. The good news is that healthy mathematics teaching can also spread like an epidemic. If one teacher or parent is "cured," the cure may spread.

When truth is fragmented into bits, it loses its core. Mathematical truth must be perceived as a whole, at least some of the time. The whole is, in some sense, more than the sum of splintered parts.[13] Learning math requires examining each part in detail but then putting them together in our minds in ever-larger clumps. Our overview must continually gobble up the details.

A recent development in educational psychology has been the concept of "right brain" and "left brain." Traditional mathematical education, including, but not only, Drill and Kill, has focused on sequential learning, one step after another. In the modern jargon, this taps into the left side of our brain. However, we also need holistic perceptions, such as those appearing in geometry and art. This uses the right side of our brain. We all need to use both sides, but some of us have developed one side preferentially. We all need to be less fragmented.

Computers are another innovation of the past few decades. Some computer programs are remarkably stimulating to the mind, but as a group they can't hold a candle to well-prepared teachers and parents. Computer scientist Dr. Michael Fellows worries about the tendency to "fetishize computers," and claims that their main beneficiaries have been the computer companies and the educators who get generous grants to figure out how to use computers. He writes, "It is quite possible that the Golly-Gee-Whiz-Look-What-Computers-Can-Do school of mathematical pedagogy will eventually come to be regarded as a disaster of the same magnitude as the 'new math' rage of the 1960s. . . . The public needs to understand that math and computer science are not about computers, in much the same way that cooking is not about stoves and chemistry is not about glassware."[14]

No computer can teach on-site communication, friendship, and collaboration. These are essential not only for world peace but also for mathematical achievement. It is nice if you can afford a computer, but don't think it substitutes for teaching or parenting. If you can't afford a computer, you still can encourage and teach your child to enjoy and know mathematics. We need to be connected to the people around us when learning mathematics.

Summary

Twentieth-century life demands mathematical power. Yes, computers and calculators can do much of the boring "math," but only human thought can decide what is important and what needs to be done. The existence of computers has made math power even more important for people, both for national prosperity and for individual egos.

Parents make the difference. During the preschool years and those essential early school years, parents can empower their children to succeed at later mathematics. Any parent can do it. *You* can do it.

The Flower:
What Is Mathematical Ability?
Where Does It Come From?

2

. . . most important of all, . . . too many people in the United States think that the primary determinant of success and learning is either IQ or family circumstances instead of effort. I don't believe that, and I don't think any of the research supports that.

—President Bill Clinton[1]

What do we mean by someone being "good at math"? What qualities lead to high math grades? What characteristics do all professional mathematicians share? Since the remainder of the book will help parents foster qualities that help children become good at math (and warn against practices that damage or destroy math ability), it seems appropriate to consider these questions first. Math-powerful people have at least five characteristics in common: enthusiasm for math, *ganas* or persistence (for which one needs hope, patience, productive skepticism, a willingness to begin over, concentration, and tolerance of frustration and failure), metathinking, a tolerance of ambiguity, and a propensity to seek similarities among differences.

Joy in Mathematics

People who are good at math have enthusiasm for it. They like it at least enough to keep at it. They don't look at a math problem and automatically say to themselves, "I hate this" or "Why bother with such nonsense?" They may be negative occasionally (we all are at times), but not routinely. Of course, it's the old question of which comes first—the chicken or the egg. Do people like math because they are good at it? Or are they good at it because they like it?

Once I commented to my college music professor Peter Graham-Swing that I didn't think I had musical talent. Without missing a beat, he replied, "I think that musical talent is loving music." I have often reflected upon the implications of that statement. Is "loving it" indeed the main ingredient of musical, mathematical, figure skating, bird identification, or car repair talent? What can be more important? Is joy a necessary, if not sufficient, condition for talent?

People who are *very* good at mathematics really enjoy doing math. People who are good enough to use it for ulterior purposes like it enough. They don't dislike it.

Ganas

Star math teacher Jaime Escalante popularized the Spanish word *"ganas"* as the quality needed for excellence in mathematics. Roughly translated it means, "urge," or "desire," or "inner drive." It has no exact English translation, but even monolingual mathematicians can recognize its aptness.

Escalante has his classes chant, "Determination plus hard work plus concentration equals success, which equals *ganas*." It sounds like a good formula to me. Let's explore it more deeply.

Jaime Escalante—Miracle Worker

Jaime Escalante, a Bolivian immigrant, was not allowed to teach in the U.S. until he repeated all his college education here. When he finally became a teacher, it was in East Los Angeles, where 80% of the annual family incomes were less than $15,000 in 1987. Many were much less.

In 1987, 129 students from his high school took the calculus Advanced Placement (AP) examination, an advanced math exam given by the Educational Testing Service (ETS), and 78 passed. This was the eighth highest number of passes of all the public schools in the entire country. His high school produced over a quarter of all the Mexican Americans in the country who passed that year.[2]

Jaime Escalante is the main character of the popular movie *Stand and Deliver*. It culminates in the drama of 1982, when 18 students took the calculus AP exam, and ETS suspected that 14 had cheated. Eventually, 12 retook the exam with ETS proctors—after they had had three months to forget the material!—and all passed.[3]

PERSISTENCE (DETERMINATION PLUS HARD WORK)

One characteristic common to all people who are good at math is persistence. People good at math don't give up easily. Real math is hard. Nobody "gets it" in an instant.

Mistakes of great mathematicians: Andrew Wiles spent seven years working alone in an attic room of his home on one math problem. Fermat's Last Theorem was first stated in 1637. Fermat wrote it in the margin of a book, adding that he had a marvelous proof but it was too long to fit into the margin. For over three centuries people have wished he had a larger margin. Many tried to prove the theorem either true or false, and books were written about the problem.

Born in 1953, Andrew Wiles first saw Fermat's Last Theorem when he was ten years old. Throughout his teens he tried to prove it, and then he decided to become a professional mathematician. When he was 33, he was able to give up everything else (except some teaching at Princeton University and his wife and two young daughters) to concentrate on "FLT." For seven years he struggled.

On June 23, 1993, he culminated three tension-filled talks in deep mathematics at Cambridge University with the amazing statement, "And, so, Fermat's Last Theorem is true." His colleagues at Princeton University got up at 5:30 A.M. to turn on their email. They read, "Wiles proved FLT," relayed from friends in England.

The following day the news appeared on the front page of the *New York Times*. For months six people studied the proof, finding small errors that Dr. Wiles then corrected. It was said there were only 100 people in the world who could understand his reasoning.

In December, some newspapers reported that Wiles had been working intensely for some time to fill a "hole" that his reviewers had found in the proof. He seemed stuck. For several months he struggled, and then he feared the rest of his life might be consumed unsuccessfully trying to fill the gap. Not wanting to share the fame of proving FLT with another senior mathematician, he invited a former student to work with him.

He discussed the problem extensively with Dr. Richard Taylor, already tenured at Cambridge University. That enabled Dr. Wiles to see the connections needed to complete the proof. Some people therefore concluded that Andrew Wiles is the greatest twentieth-century mathematician.[4]

Everybody makes mistakes, and we need others to correct and help us. Most people don't have the ability to work eight years on a single problem. Most of us wouldn't have the financial resources! Dr. Wiles, however, gained those financial resources because of his unusual persistence and *ganas*. At age 33, he was already recognized as a top mathematician. His employer was willing to bet his research salary that seven years of *his* persistence would produce worthwhile mathematics.

What contributes to a person's ability to persist? Six necessary qualities are hope, patience, productive skepticism, the ability to start over, concentration, and tolerance of frustration.

Hope: Determination requires hope.

If you don't try, you won't succeed.

If you don't knock, it won't be opened unto you.

If you are convinced you can't do something, then you probably can't.

New mathematical understanding occurs after we focus on a problem or concept and wonder about it. The problem can be self-generated or posed from without. In either case, our first reaction may be one of frustration. It is new, and we don't see how to begin. It seems strange. We are blocked.

Hope must whisper, "I can do this. There is something outside my reach that I don't yet have. I *will* get it."

The struggler tries to snatch the answer. It is elusive. The struggler tries again. And fails. The struggler tries another approach. And fails again.

Trial and failure. Trial and failure. I experience it myself, and I watch it in the youngest preschooler. I see it in elementary school children and in successful high school and college students. Failure precedes success. To acquire mathematical understanding, one must have the courage to risk failure and the emotional stamina to pick one's self up again and again after failing. This requires hope.

Patience: Sometimes to solve a math problem, you have to sit and wait for the correct approach to hit you. You have to concentrate patiently before you perceive the relationship that makes the solution possible. On the outside, it looks like nothing is happening. Sometimes it feels that way on the inside too.

Finding the "path" to solve a math problem requires finding the beginning of that path, and possibly chopping it out of the thicket. This requires patience. To wait without turning off, you must believe that the answer may come. You must hope. You must be patient.

After you have found an approach, patience is essential for the actual solution of long math problems.

Productive skepticism and a willingness to start over again: Sometimes your original path doesn't lead where you intended. When you persist to its end, the answer doesn't check. It is wrong.

There are two possibilities: the method may be wrong, or you may have made a mistake. First you will want to retrace your steps. Did you do what you intended to do? If you can't find a mistake, you must go back to the beginning and contemplate an entirely new approach.

Because we so often try the wrong method first, "doubt-that-this-is-a-worthwhile-path" is essential for challenging mathematics. We have to be able to be skeptical of what we are doing, question its merit, and turn back. We have to be able to give up and try another path. We also need to be able to catch our errors as we move along a planned path.

In math it is more obvious when you are wrong than in most pursuits. Answers that don't check are wrong. Approaches that don't seem to lead anywhere are suspect. To spend much time doing mathematics satisfactorily, you must be able to say (at least to yourself), "I was wrong. I will try another way." This hurts a bit, but probably not as much as the consequences of not being able to say it.

Concentration: Anyone who has watched a baby examining dust on a rug knows that young humans have considerable ability to concentrate.

As children get older, their absorption in a single activity can be irksome to those nearby. Anyone who has watched a child in a store repeatedly asking, "Why, Mommy?" when it is obvious that Mommy is trying to ignore the question has another glimpse of youthful concentration. Preserving the inborn ability to concentrate, so essential to solving difficult math problems, can be challenging for a parent, but it is worth the effort.

Tolerance of frustration and failure is key to patience and persistence. Tolerance of frustration means you can't be too passive. You keep trying other paths. You keep fighting the intellectual darkness. You keep believing there is a better way.

Could your child tolerate frustration? Yes, indeed. Kids come to us with an amazing level of frustration tolerance.

Watch a two-month-old in a baby seat studying his or her own hands. Waving those fascinating hands about, the infant discovers a connection between them and "what I do." Can I make them go where I want them? How? Watching an infant learn to control her or his own hands raises one's opinion of the human race. Such persistence! Such curiosity about the powers that I have!

A few months later come the experiments with a food-filled spoon. Splat! Food decorates the small forehead. Try again. Up the nose. Try again. An ear. Try again. The mouth! Food in the mouth! Triumph for parent and child.

Remember the creeper trying to mount the stairs. Each stair looks too tall for such small legs. But the arms strain and stretch and finally pull the body up onto the next stair. Sometimes the body tumbles back to the floor. Failure happens. Success may follow.

A parent suggests, "Bring me two spoons." Another attempt. Another failure.

"Yes, very good. That is a *spoon*. But I want *two* spoons. Can you bring me another spoon?" Too soon. The child can't bring two spoons. Oh, well, a cuddle and a giggle will do.

But then one day the child does bring two spoons. "Very good! That's two spoons! One, two!" The delight in the caretaker's voice is distinguishable from the earlier tries, even if the cuddle and giggle are similar.

It may be just an accident that there were two spoons. The child may not have made the association between the word "two" and bringing two. But eventually he will, just as eventually the hands will go where the two-month-old commands, and the spoon will go in the mouth, and the legs will go up the stairs.

Toleration of frustration and failure is inspiring among young humans. Folks who cheer children's success supplement their inner rewards. The children gradually gain control over their body and environment. If families try, mathematics can help teach frustration tolerance. Success is frequent *enough*, and failure is safe.

Internal Scripts and Metathinking

Experienced mathematicians have familiar scripts for an internal dialogue that can be learned and taught by math neophytes. They include:

- ► Can I do an easier problem?
- ► Does this problem remind me of another problem I can do or almost do?
- ► What *does* this remind me of?
- ► Can I alter this problem to become a related one I have solved?
- ► Can I alter this problem and solve the related one?
- ► Can I do a piece of this problem?
- ► Is this problem really another similar problem in disguise?
- ► Can I draw a picture describing this problem? A graph?
- ► Is this problem a special case of a pattern I already understand?

These scripts don't work automatically. Part of developing mathematical ability is internalizing the questions and keeping them "well greased." Parents can try to learn them and share them with their children. People who are good at mathematics ask themselves such questions as naturally as most Americans ask, "How much does it cost?"

Math-prone folk continually seek relationships between the known and the unknown. We are always trying to apply what we know in one context to another. We are always looking for ways to avoid memorization.

When we are not deeply engrossed in some activity, we tend to "metathink." In other words, we think about our own thinking.

- ▸ Is this what I should be thinking about?
- ▸ Is there another way to go about this?
- ▸ Is there an easier way?
- ▸ Does this make sense?
- ▸ And then what?

So goes the internal dialogue of a mathematician. You can foster it in your child and yourself.

There are four types of questions when "doing" mathematics. They flit through the mind of someone solving a specific problem or simply exploring ideas.

1. Are there other approaches? Am I barking up the wrong tree? Can I do it another way?

2. Is this part of a bigger problem that is easier to tackle? What are the generalizations?

3. Is there a smaller part of this I can solve? Or a similar simpler problem? Will that give me a clue how to solve the problem I am working on? Does it remind me of an easier problem that I know how to solve? Can I reduce this problem to a known problem? Can I work out a particular example?

4. Is this the right problem? Or is it really a different problem that I am trying to solve? How often do school children grapple with the wrong problem in the sense that they misunderstand what is being asked? Often!

We can summarize these basic mathematical questions:

1. Another way?—parallel
2. Bigger problem?—up
3. Smaller problem?—down
4. Right problem?—Scrap and start again.

Mathematicians tend to look at life this way. We are metathinkers. Sometimes we sink deeply into a problem so that it absorbs our being. But when we come up for "air," we monitor our thinking and wonder if it should be redirected. We ask ourselves our menu of internal questions.

Tolerance of Ambiguity

"Ambiguous" statements are "capable of being understood in more than one way."[5] Despite math's reputation for preciseness, ambiguity can be regarded as the bedrock of mathematics. Perceiving a pattern means recognizing how two things are the same although they are different.

- ▸ Three wheels on a tricycle
- ▸ Three corners on a triangle
- ▸ Three leaves on a clover
- ▸ Three loops on a pretzel

How are they the same? How are they different?

Math ability requires tolerance of ambiguity. What was "hard" becomes "easy." Math demands logic, but intuition is also vital. Math is full of psychic balances. To grow mathematically, we must try to be:

- ▸ self-confident, but modest
- ▸ hopeful, but aware that failure is unavoidable and continual
- ▸ disciplined, but joyful and creative
- ▸ able to focus, but also able to enjoy aimless play
- ▸ alert to external stimuli, but tuned in to messages from our inner self
- ▸ able to pay attention to another person, but also able to reconsider the value of the task at hand

Several of these paradoxes reflect the need to concentrate without "losing" oneself. Some obsession is needed, but not too much. Your mind must be open to new ideas, but closed to distractions. How a "new idea" differs from a "distraction" is a very gray area.

Recognizing Similarities Among Differences

Since math itself is about patterns, math ability is ultimately the capacity to make connections among seemingly disparate concepts, arrangements, and objects. Math is perceiving similarities among differences, noticing the same pattern in different guises.

Research mathematicians are expected to make mathematical connections that nobody thought of before and prove them. They thereby notice new patterns and name them. If you took high school geometry, you proved theorems that others have been proving for

Combinations

All of these have the same pattern:

- ▶ The number of boys in a family of four
- ▶ The number of heads in a toss of four coins
- ▶ The number of blue Unifix blocks in stacks of four blue or white Unifix blocks
- ▶ The number of acceptable items among four products made by a machine that makes half of its products correctly
- ▶ The number of odd digits among any four digits in a randomly generated number

The Four-Color Problem

In 1852 a young Englishman wrote a letter to a student in London, asking if he could find out whether four colors were enough to color any map without any two adjacent countries having the same color. It is easy to see that the map below requires four colors.

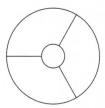

One of the leading mathematicians of the time, Augustus DeMorgan (1806–1871) soon proved that five colors were enough for any map. But the "Four-Color Problem" remained unanswered until 1976, when Kenneth Appel and Wolfgang Haken reduced it to a long-running computer program. A computer at the University of Illinois at Urbana ran all night and decreed in the morning that, yes, four colors are enough.

No human has understood this proof. It is too long and complicated. We don't know whether an easier proof exists, but some mathematicians have checked the reasoning of Appel and Haken. Most mathematicians believe that four colors are enough to color any map—although thus far only a computer has "understood" the entire proof![6]

over 2,000 years. Generally you used the same proofs that students have used over those centuries. When you were imaginative, your teacher may have chastised you as "wrong," even when you had found another right proof. (Perhaps, however, your teacher praised your creativity. I hope so.) Research mathematicians are creative, making connections and noticing patterns in increasingly innovative ways.

Mathematical ability includes the enthusiasm, the urge, the confidence, and the persistence to investigate patterns and use them to solve problems. It involves some habitual internal scripts, habitual metathinking, a tolerance for ambiguity, and a propensity to make connections. And, yes, there is something more. What makes those surprising connections? What causes *ganas*? Something in human ability can't be described or analyzed. It's miraculous. But mathematical miracles happen often when people believe they are possible.

Where Does Mathematical Ability Come From?

I believe we were all conceived as math geniuses. It's downhill from there. Perhaps both parents' diet and drugs (prescribed or otherwise) before conception make a difference; certainly a mother's prenatal diet and drugs matter. Birth itself is extremely hazardous; I personally know people whose mathematical capability was damaged by a difficult birth.

At any time, a head injury, bungled brain surgery, oxygen deprivation, or meningitis can decrease ability. Diet and drugs continue to be hazards. Our ability to solve math problems is affected by physical and emotional traumas, often permanently.

After birth, nourishing a child's natural talent and protecting it from the slings and arrows of outrageous fortune makes a huge difference. This book addresses what parents can do directly, both to nourish and to protect. It also discusses how parents can affect their children's teachers' impact.

Short-term, math ability is affected by lack of sleep, lack of exercise, emotional distractions, and noise. One student came to my office frustrated that she couldn't get the test grades that she and I thought she deserved. After listening to her for a while, I suggested she go for a brisk walk for a half hour before each test. Her grades promptly rose from a tottering C⁻ to a steady B⁺. Blood circulation affects how you think. Turn off the TV and send your child out for old-fashioned outdoor running-around play.

Asserting that environment has enormous impact says nothing substantial about the effect of heredity. Thomas Edison presumably said, "Genius is one percent inspiration and ninety-nine percent perspiration." Albert Einstein allegedly modified this to "Genius is ten percent inspiration and ninety percent perspiration." Michael Pollan reports that "Genius is five percent inspiration and ninety-five percent perspiration," and then carries on about how important that five percent is.[7] I agree. Something beyond diligence is needed for creativity. I don't know what, but I think it may constitute more than the one percent that Edison reputedly claimed.

Personally, however, I suspect that *ganas* is more important than genes even in that unexplained portion of genius. I believe we all have some free will. Can a two-month-old decide whether to pay attention or cry? Can a seven-year-old decide to defy our anti-intellectual culture and learn as much as possible? Can a teenager decide, despite incredible obstacles, that the AP calculus exam is conquerable? Can a parent decide to impart an urge to explore patterns and make connections? Can a parent allow persistence in children? Can parents stimulate independent thinking in their offspring without frustrating themselves too much or abandoning their parental responsibilities?

None of these questions has a clear answer, but I suspect that what an individual decides does make a difference. I happen to know a pair of identical twins, now in third grade, one of whom loves school. The other has been sufficiently unhappy there that the mother has decided to try home schooling. Same genes. Same loving home. Extremely different personalities. *Ganas?*

How else do we explain Mary Fairfax Somerville (1780–1872)? Her father forbade her to study the algebra and geometry books her brother's tutor had smuggled to her, but she read them in secret. Not until her second husband, Dr. William Somerville, encouraged her mathematical activities could she pursue them openly. When she did, she became one of the leading mathematical writers of her time.[8]

Even more extreme is the story of Sophie Germain (1776–1831), whose parents discovered her studying math at night and then took away her fire, her light, and her clothes in an effort to stop her. She stowed away candles and used them to study, wrapped in blankets after her parents had fallen asleep. As an adult, she initiated a correspondence with Gauss (mentioned on p. 179) using a fictitious male name, and he praised her originality in mathematics. Later she won a competitive prize for research mathematics, judged anonymously.[9]

Ganas is inexplicable. *Ganas,* parents, and teachers seem far more influential than heredity to me. I believe there is no innate difference among the races or between the sexes in math ability; there is no convincing evidence that there is. Even if individuals begin with the same math potential, by kindergarten this world has already had an enormous impact on every child.

Despite parents' considerable influence on their offspring, plenty of other factors affect a child. Parents don't control all their child's circumstances, even when the child is very young. Parents have even less power later. However, they are not impotent, and parents who follow the guidelines in this book tend to raise children who are "smart at math."

How do you know if you are successful? Getting good grades on math tests or high scores on standardized tests is a good sign. Enjoying math may be an even better omen. Thinking up clever problems and answers is better yet. However, there is no sure way to tell.

> How late is too late? One of my best graduate students really "wowed" his (excellent) class-mates and me with his impressive knowledge and wide mathematical reading. One day in my office he confessed that he had never learned what fractions were until he was married. In the privacy of the marriage relationship, his wife taught him this exciting subject, and he went on to study math avidly. When he left me, he was scheduled to teach calculus the following fall.

Understanding mathematics is complex. It is a repeated leaping of the imagination to ever-new heights. It is an awareness, an appreciation, a flicker of ideas in one's mind. It is a mental openness that can be stimulated and encouraged, but there is a miraculous aspect that rational intentions cannot touch.

Mathematical insight comes. We must have faith that it will come. We must wait for it. We must encourage it. But we can't force it, even in ourselves. Miracles cannot be scheduled.

Even more, we cannot force mathematical understanding into children. They must want it, prepare for it, and wait for it. They must rejoice in the coming of mathematical vision. Little children do. They are diligent, eager, and joyful about understanding math. Our job is to cultivate their budding native ability and enjoy its flowering.

How Math Blooms 3

. . . you are surprised at my working simultaneously in literature and in mathematics. Many people who have never had occasion to learn what mathematics is confuse it with arithmetic and consider it a dry and arid science. In actual fact it is the science which demands the utmost imagination . . . it is impossible to be a mathematician without also being a poet in spirit. It goes without saying that to understand the truth of this statement one must repudiate the old prejudice by which poets are supposed to fabricate what does not exist, and that imagination is the same as "making things up." It seems to me that the poet must see what others do not see, must see more deeply than other people. And the mathematician must do the same.

—*Sofia Kovalevskaya, 1890*[1]

Mathematics is different from other subjects. It really is. It doesn't depend on world view. The math of ancient China wasn't identical to that of faraway Rome, but they were compatible. One culture may have missed something that the other included, but when it was discovered, both cultures agreed it was right.

Math is the only subject in our elementary school curriculum that would be exactly the same if there were thinking beings on another planet in another solar system. Their history, language, and biology would be different from ours. But their math would be the same. Math doesn't depend on what we see, hear, smell, taste, or feel.

One curious result of this universality is that no matter how we learn math, we gain the same insights. Mathematics is the same, worldwide and universewide. We can learn math in many different ways, but we always come to the same conclusions.

The International Congress of Mathematicians meets every four years. After they listen to each other's proofs, mathematicians from all countries agree on what is right! There will always be many things we, collectively, don't know. We enjoy sharing what we have recently learned and "putting our heads together" to solve currently unsolved problems.

MATH IS THE ONLY SUBJECT THAT WOULD BE EXACTLY THE SAME ON ANOTHER PLANET.

But the solved problems are solved to the satisfaction of everyone who understands the question and answer. (Oh, yes, sometimes mathematicians make mistakes. Everybody does. Occasionally these mistakes are well publicized, as Andrew Wiles' was. But mistakes are mistakes, and everyone agrees. Usually they can be remedied, and when they are, everyone agrees again.)

The flip side of the fact that mathematical truth is the same no matter how you learn it is curious indeed. Children taught in lockstep tend to forget what they have learned or remember only fragments. *The upshot is that diverse learning approaches yield uniform results, and rigid teaching yields varying (wrong!) results.*

The public has trouble with this. It doesn't seem right. Shouldn't all kids have the same opportunities? No, for two reasons:

1. Kids are different from each other.
2. The adults around the kids are different from each other.

Efforts to force a uniform methodology tend to backfire because of these two facts. However, some standard achievement expectations over a period of years *are* realistic. If a teacher understands the mathematics to be learned, knows a variety of ways of enticing the children, and has the time and interest to pay attention to each child's efforts, then the children will learn. It happens in other countries, and it can happen here. If it's not happening in your local schools, you may be able to make a difference. You certainly can affect what happens at home.

First, your home math activities must be pleasant. Nothing is worth violating that principle! Secondly, there are at least two types of ways of learning math. One—practice and rote—is well known. However, the second—tapping the inner life of the individual—is more successful. Significant math is learned in Leaps of understanding. A broader picture suddenly becomes clear. Psychologists may call this the "Aha" experience. Educators have dubbed it "the having of Wonderful Ideas." More about them later.

Biking and Exaltation

Learning math well is more like learning to ride a bike than increasing skill at running. If you want your child to cover ground quickly and go far, you can coach techniques of cross-country running. You can insist upon daily practice. Your child will develop speed and stamina. Any healthy child can run faster and farther with coaching and practice, but there are limits.

If you provide a bike and let your child try to ride, there will be many spills. You won't see daily improvement. A friend who is running will cover much more ground. Your child will experience repeated frustration and failure. Different techniques must be tried, techniques that are strange and new to your child. You won't find them easy to explain.

But one day, your child will hit upon the right balance, and suddenly that bike will go forward. At first it will be a bit wobbly. The child may still have an occasional spill. But soon bike riding will be almost as easy as running.

When that happens both the speed and the distance covered will be far greater than that possible by running. A whole new vista of achievement opens up. Now daily practice and coaching in biking techniques will result in much faster speed and longer distances. A new level of practice is merited and new standards will be achieved. No matter how much runners practice, they will never attain the speed of children who have tolerated frustration enough to learn to ride a bike, even though at first they moved not at all.

Such radical, sudden changes in expertise occur repeatedly in the process of learning math. Some practice and coaching is desirable at each stage. But more important, a learner must be allowed to struggle to Leap to another level. This struggle requires time, space, and satisfaction with past achievements.

Others can provide a metaphorical "bike," the alone time to experiment with balance, and assurance that the quest is worthy. Others can reiterate that failure is merely a prelude to success. They can provide a joyful cheering section. But only the person riding the bike can discover balance. Ultimately, we each learn alone.

Too often people in positions of authority force children to "run" faster and longer instead of allowing them time and emotional space to attempt to ride a bike. Children want to gain new balance, to see a larger picture. Only the child's own investigations will enable her or him to balance a bike.

What is this miracle of balance? How does it happen? Much of this book is devoted to exploring how parents can provide a similar "balance" in mathematics: through games, manipulatives, conversations, questions, allowances, and other activities that help a youngster into the next level of mathematical understanding. At each stage there is another "bike" to be balanced. First, however, let us look more closely at another aspect of the similarity between learning math and learning biking.

Biking and Humiliation

Once you learn to ride a bike, it is easy. You wonder why you had trouble before. How could I have been so stupid last week? Why didn't I do it before?

This may be the most irritating aspect of learning mathematics. Once you make a new Leap and get a new level of understanding, it seems very easy—so easy you wonder why you were so stupid not to see it before. With just a little encouragement, you beat on yourself unmercifully. "I must be very stupid to have learned this so slowly. It is so easy. It took me so long." Perhaps this repeated humiliation is the reason why some people hate math. It's worse in math than with learning to ride a bike. Self-deprecation about our previous inability to ride a bike is countered by others who notice and praise our new skill. But mathematical progress may not be noticed by others. Who is to counter the inner voice telling us we are stupid because we didn't learn it *last* week?

This is where parents are vital. A busy teacher with dozens of children barely has time for a "Good!" Watching some teachers exclaim "Excellent" for forty solid minutes in an effort to affirm their pupils' progress can be wearing to the point of disingenuousness to me, and probably to the recipients of their nonstop praise as well. (Teachers who are never pleased, of course, are worse for their pupils' psyches.)

Parents can notice and be pleased. Your style in showing your pleasure is *your* style, but you should try to observe your child's progress well enough to be aware of and rejoice in her or his individual triumphs. Never deprecate them! Each intellectual achievement takes effort for the person attaining it. A parent's most important role in their child's intellectual growth is to notice and enjoy successes.

A child who learns that repeated failed attempts often end in success has learned one of life's greatest lessons. That may be *the* clue to raising a child who will achieve. We all learn that failure is inevitable. An infant knows that failure is sometimes followed by success. Watch one trying to turn over. Or a baby who is determined to walk.

Failure after failure is followed by success, only to be followed by yet another fall. Full of innate hope, a young child continues. What we adults need to do is to refrain from destroying that inborn belief that failure is a precursor to success and that persistence is worth the effort. You must have a resilient ego to live with yourself while learning math.

Again and again, I have watched this self-anger at previous stupidity in preschool children, in graduate students, in research mathematicians, and in everyone between. It's the worst part of learning mathematics for me still. I keep worrying that people are going to find me out. "How slow I was learning that new concept! It is so easy compared to other mathematics I learned before. Perhaps I am losing my grip." Decade after decade, I harbor the persistent suspicion that I am losing my grip.

Mathematician Hugo Rossi suggested that perhaps the greatest service mathematicians could provide to public education would be to confess openly how difficult we too find math. I hereby confess. "The discomfort of intellectual uncertainty is very hard to tolerate, all the more if you are on the outside looking in," wrote Rossi.[2] The ability to tolerate "intellectual uncertainty" is fundamental to mathematical growth. Ambiguity and frustration are part of the experience. If you feel "on the outside looking in," it is worse. You want your child to feel on the inside. Frustration is normal. Understanding takes time.

Furthermore, children begin school sure they will learn. It is wonderful to see little children's faith in school, even in the most distressing neighborhoods of our country. They know school is where the big kids are, and where they too will grow bigger.

This faith usually continues through third grade. Parents and teachers who want children to continue to tolerate frustration and failure beyond that must provide appropriate challenges—not so dangerous or difficult that they will damage the child, but hard enough so the child gets a real sense of achievement by meeting the challenge. Confidence comes by experiencing success after failure. "Confidence is the knowledge that one can achieve one's goals; it is a faith based on achieving something tangible," writes Charles Sykes.[3] The National Association for the Education of Young Children agrees.[4] Nothing succeeds like success. When we succeed, we want to continue.

Meadows and Brooks

In real mathematical insight, as in biking, you don't forget. You can stay away from a bike for years, but once you have learned how, you can do it again. You may be hesitant at first, and some practice helps, but soon you are back nearly to your old speed.

The same is almost true with mathematics. When you absorb a new concept, you don't forget. You may need your mind jogged a bit, and you certainly will need some practice to get up to speed. But the concept comes back much faster than the first time you learned it—if you really learned it the first time.

A skeptic may rightly question what is "really learned." You may have noticed that your ability to forget mathematics exceeds your ability to ride bikes. Language is treacherous, so I turn again to another metaphor. Please bear with me. Remember that because mathematics cannot be seen, heard, tasted, smelled, or felt, it is not easy to describe. Religious leaders, when discussing similarly elusive topics, turn to parables, myths, and images. One problem with the biking metaphor is that you learn to ride a bike only once. You may get a better bike, but your increased speed and distance are marginal. If you get a motorcycle, it uses more than your own skill and stamina. And what then? Your private plane? And then?

In mathematics, the analogous experience to learning to ride a bike happens again and again. A professional mathematician experiences it repeatedly for a lifetime. A well-prepared citizen would experience it much more frequently than the average graduate of today's U.S. high schools. Let us explore another image of how math is learned.

THE MEADOWS-AND-BROOKS IMAGE OF MATHLAND

Suppose mathematical truth were a landscape consisting of Meadows and Brooks. The Brooks separate the meadows from each other. (This is not realistic, since each math land Meadow is, in effect, an island, and real-world islands are not separated from each other by brooks. But this image makes the language easier.)

If you are in one Meadow, it is easy to wander around the rest of it. But to get to another Meadow, you must Leap across a Brook. Leaping is not easy. It involves intention, and courage, and a bit of luck.

Learning mathematics involves many Leaps over Brooks interspersed with exploring Meadows. Exploring the Meadows is relatively easy compared to jumping the Brooks. The next chapter shows one way of interpreting the exploration of numbers using this Meadows-and-Brooks metaphor.

Alas, our educational system too often pressures children to stay in one Meadow. There is too little encouragement—or even emotional space—to gather the joy, courage, and vision to Leap to another Meadow.

It is true that exploring each Meadow is justified. Indeed, this may be the most time-consuming activity in learning mathematics. Adults can help by providing exercises and maps. Teachers can plan detailed lessons with a "participatory set," and "closure."

But it is the Brook-Leaping that yields the most dramatic short-term breakthroughs and durable long-term memories. For Leaping, the learner must be the initiator. Teachers and parents can provide manipulatives, games, conversation, and encouragement. They can pose pregnant problems. But the child must solve the problems in *the child's own way at the child's own pace.* Only slow, careful investigation will enable the child to take another "Leap of Understanding."

Plodding Progress and Leaps

Allowing Leaps of understanding is risky business but many teachers do it effectively. Children do learn. In *Garbage Pizza, Patchwork Quilts, and Math Magic,*[5] Susan Ohanian tells about many different projects sponsored by the Exxon Education Foundation that dramatically increased the mathematical standards in dozens of localities around the country. Individuals like Marva Collins,[6] Robert Moses,[7] Jaime Escalante,[8] and Seymour Spiegel have performed miracles in some of the country's least promising neighborhoods using sheer knowledge, ability, caring, and incredible personal energy. In all these cases the teachers provide a fertile learning environment.

Timing is crucial. Effective teachers (in the classroom and in the home) *listen* to their students so they can understand where they are—on which Meadow they are now functioning. Memorizing sums is much easier after you understand their purpose and power. If you don't understand the concept of 5, +, and 8, then 5 + 8 = 13 is meaningless gibberish. You would rather memorize rhymes about the cat and the fiddle.

This is why children must be on the addition Meadow before memorizing addition facts. Until then, they need more imaginative activities to entice them to explore simpler Meadows and jump over Brooks. Once a child understands the purpose and power of sums and differences, they are quickly memorized. If a child understands the concept of plus and the addition facts, finding the number of marks in the pattern below is much faster than by counting.

```
        X       X           X   X   X

                X               X       X

        X       X           X   X   X
```

Most children in this country eventually arrive in the addition Meadow and then learn their addition facts. The tragedy is that so many don't move beyond. Too many U.S. youngsters have been stranded on the addition Meadow. Perhaps they never see the vision of another Meadow beyond the Brooks that trap them. Or maybe they aren't given the time and safety to Leap. Or they are so beaten down, they don't have the intellectual courage to Leap.

Aliens

If someone gives you a worksheet from a Meadow beyond your comprehension, you don't understand the language. You are an alien in a strange land, a misfit who wants to hurry home. You don't learn anything because you aren't really there. At best, your body goes through the motions. Your spirit is peering into a world with no sense of belonging. You are an alien.

Aliens are anxious about how they will be treated. No wonder! Just look at history! Or the inner life of anyone. When we are alienated, we are anxious. "What destroys joy in mathematics is not practice but anxiety—anxiety that one is mathematically stupid, that one does not have that special numerical talent. But math talent is no more rare than language talent." So writes E. D. Hirsch in his four books, *What Your nth Grader Needs to Know* ($n = 1, 2, 3, 4$).[9] I agree with this passage. But I disagree, vehemently, with his accompanying assertion that practice is the key to *learning* mathematics. It helps in remembering, but practice in math must follow understanding because *the facts follow from understanding*. Healthy practice of *ideas you already know* to implant them firmly or to tack down details (such as learning "basic facts") is pleasant in the sense that shooting basketball baskets between games is pleasant. It isn't as exciting as playing an actual basketball game or making the Leaps in mathematics, but practice is necessary between the peak joys of basketball and math. It keeps those peak joys coming, preparing us for the next. Practice is part of life, whatever we do. Math is no exception.

Fostering Leaps of Understanding

Continuing excellence in math involves expanding visions, seeing relationships among patterns more deeply and clearly than before. How can we help our child make those Leaps? There may be four core guidelines: joy, space, listening, and providing a setting (which I will sometimes call "gardening").

JOY

Math is supposed to bring joy. That's how it blooms. Even if you have been affected by our country's anti-math epidemic, try to hide this from your child. Make your mathematical interactions with your own child as much fun as you can, especially when your child is young. Fortunately, the time when a child is most vulnerable is also the time when you have least reason to be anxious about your own inadequacies. You can count. Count and laugh. Count and sing. Count and dance.

What could be more fun than capturing wonderful ideas? Laughter and joking must be part of the scene. The limits of human aspiration prompt a chuckle as well as a tear. The setting will have many humorous aspects, including as much laughter as you can manage. The more joyful you can keep your math sessions, the more likely your child is to succeed.

As your child poses harder and harder questions, you may find it harder and harder to remain joyful as you approach them. Remember nobody's grading *you*! Your child will thrive if you act as if there were nothing you would rather do than think about the problem at hand.

Remember that tolerating frustration is a key to mathematical achievement. Try to keep the "Anxiety Bear" at bay longer each time. Neither you nor your child can be on all Meadows at once. Perhaps you have just drawn a blank. Keep up the conversation. Ask questions. Listen carefully to the answers. Savor this sharing with your child. Say a prayer of thankfulness for the privilege of being an involved parent. Let your child know how grateful you are to be together at this moment.

The first time your child solves a problem before you do is a challenge for any parent. Can you rejoice in your child's triumph? Can you seize this precious moment? You are truly an excellent math parent! Praise your child profusely for this success. If you can keep your attitude positive, you have both Leapt across a big Brook! (One parent mused at how much easier it is to rejoice in an offspring's excellence at playing an instrument that the parent can't play than it is to feel good about a child's superior math ability.)

SPACE

Your child needs "space" to grow mathematically. Don't ever force math on a young child! Let math emerge frequently in your conversations, but when your child doesn't "catch it," don't press and don't worry. All in good time. Joy is the key to long-term learning. Too often, in our homes and schools, we stifle miracles by pressuring them to come too soon.

Mary Leonhardt claims that a major reason so many youngsters today don't love reading is our propensity to classify them as "low" if they don't get up to speed rapidly. Similar observations apply to math. She writes:

> Some children, by the age of eighteen months, are speaking in complete, short sentences. Other children don't acquire sentences until almost a whole year later.
>
> How do we handle this? Do we segregate the children who are slow in speaking? Do we spend hours a day drilling them? ("Say *chair*. Now again!") Of course not. They play all the time with their more fluent friends, we talk to them all the time, and by the age of three or so they've caught up. By the time they're four no one even remembers which child started talking first.[10]

Children need space. "Children of all ages need uninterrupted periods of time to become involved, investigate, select, and persist at activities," says the National Association for the Education of Young Children.[11] (Adults need such time too, a fact too often ignored by our frenetic society.) Overplanning by zealous parents and teachers can smother. Take time to stand back and watch. Your belief that your child will be able to Leap will greatly aid him or her to do so.

LISTENING

Eleanor Duckworth's "The Having of Wonderful Ideas," was an influential paper in 1972 about intellectual Leaps. She described the joy and amazement of children who ask their own questions and then answer them. The paper provides experimental proof that children who are allowed to find their own exciting problems and solve them at their own pace achieve the most intellectually. Allowing the "unexpected" is one of Duckworth's major themes.

> There are two aspects to providing occasions for wonderful ideas, then. One is being prepared to accept children's ideas. The other is providing a setting which suggests wonderful ideas to children—different ideas to different children—as they get caught up in intellectual problems that are real to them.[12]

Duckworth's "accepting children's ideas" is close to what I call "listening." It is a much more complicated process than either Duckworth or I can convey. Minimally, it includes showing interest, appreciation, and respect.

There is something more. We grow best when someone really cares. It is hard to define what caring is and impossible to tell someone how to convey it.[13] However, children learn best in the presence of parents and teachers who care that they learn. Partly because of the inevitable humiliation that accompanies learning math, we need a cheering section. Parents are designed to cheer.

The movement for "active listening" advocates two types of response when you mainly want to *hear* your partner, rather than interact. One is to **restate** whatever the other person said in your own words, and the other is to **articulate verbally** the feelings you hear being expressed.

Restating ideas: "Yes, I too wonder how many people are around this table" is a math-building response to a question from a child. It doesn't answer the question, but encourages the child to find the answer.

"That's the answer I got too!" is always a reassuring response to a child's guess.

"Tell me how you got that" is far more supportive than "That's wrong." Allow the child to speak as much as possible in a math setting. Don't tell the child any more than you absolutely can't resist saying. Retention is far more likely when children discover (or invent?) math themselves.

Remember that the accuracy of any particular answer rarely matters much. If you are sharing popcorn and the child who gets less will complain, then maybe it is important to get the "right" answer. But most of the time, don't sweat it. If a child doesn't see the answer or has lost interest, drop it.

Articulating emotions: "Isn't it fun to count people?" is a nice response to a child's question about the number of people sitting around a table. As often as you can, rejoice in the pleasure that you and your child are feeling with math. This is especially important where schools emphasize the opposite.

Admittedly, not all emotions will be immediately joyous. When a child is frustrated with a math problem, there's no harm in saying gently, "You are finding that hard" or "Wow, that's a frustrating problem!" Just don't imply that the difficulty of math makes it unpleasant or impossible—or that the child is in any sense inadequate.

"That will make you feel really good when you find the answer!" should be the message of your sympathetic statements.

Don't hazard a guess as to whether the child can do it if you think it's not likely. You don't want to discourage your child by saying the task is impossible, nor do you want to set goals that will cause unnecessary feelings of failure. "That's a really hard problem!" is a noncommittal statement that affirms the feelings of the struggling child without implying an expectation of either success or failure.

If you listen actively, you will discover some startling aspects of your child's mathematical life.

A father read his fourth grade child a word problem, "Together, Jihad, Ruth, and Candilaria found 24 eggs. If they all found the same number of eggs, how many did each child find?"

"Let's see. That's 24 divided by 4, so each child found 6."

"Why did you divide by 4?" asked the puzzled father.

"There were four children: Together, Jihad, Ruth, and Candilaria."

The father is a college professor who was startled that his nine-year-old didn't know the meaning of "together."

Words are slippery. The more standard English words your children know, the more likely they are to score high on textbook problems and standardized math exams. We will discuss some of the most problematic words in Chapter 8. Active listening is the key to finding out whether communication is hampering your child's math.

GARDENING

Providing a setting is the focal theme of this book. For lack of a better word, I call it "gardening." How do we provide a setting that stimulates Wonderful Ideas in children? The answer is not simple.

► It is like being a gardener, who cannot force the plants to grow, but does provide a setting that affects how well they grow and bloom.

► It is like being a stagehand for a wonderful play that has not yet been written.

- It involves pointing out math situations and being alert to a child's math wondering.
- It involves talking about math ideas and facts with our child.
- It means playing games with our child as part of our family recreation.
- It means allowing space for mud art, sandboxes, and measuring tea bags.
- It means deliberately and frequently using math-crucial words and being alert as to whether our child has picked them up.
- It involves examining the rest of the garden and encouraging the rest of the village to nourish it.

Later chapters of this book will explore these topics further.

Much of the gardening involves seizing "teachable moments." Whenever we don't respond to a child's question about number, space, or planning, we lose an opportunity to help that child develop. Every time we miss an opportunity to ask an appropriate math question, an opportunity has slipped. More important, each time we press our will arbitrarily on a child, oblivious to the child's yearnings or protests, that child is less likely to Leap a Brook to another level of understanding.

But we too are human. We can't always be listening, alert, and restrained. Nor should our child expect us to be. Young humans were created to survive incredible ineptness in their elders. Good thing, too! We all err, as we too Leap over our own Brooks of understanding—or don't.

Furthermore, math is not the only thing we want to teach our youngsters. They need love, discipline, and protection. Listening to their questions, valuable as it is, sometimes must yield to other needs. We need to be patient with ourselves as well as our youngsters. With patience, however, we can keep a vision.

If you are capable of conversation with your child, you can explore ideas together. You can exchange questions, whether or not you know the answers, and respect the responses. Please don't be pushy or worried. It can be great fun. Enjoy!

The Wonderful Concept of Three and How It Changes

4

Yet one final question remains: how can you best build up an image of something you do not yet understand?

Metaphors deny distinctions between things. Problems often arise from taking structural metaphors too literally. . . .

Metaphor, I believe, is . . . essential to anyone attempting to make sense of mathematical speech or writing. . . .

The main . . . threat . . . is the very elusiveness of the metaphor. . . . If students are not used to having mathematics make sense, then having tried and failed to make literal sense of a statement, they are likely to give up, as there is nothing to indicate they are grappling with something new. This experience may even undermine their faith in the sense of situations where the original *meaning is the most sensible one.*

. . . the greatest danger is that the unexplained extension of concepts can too often result in the destruction *rather than the expansion of meaning.*

—David Pimm[1]

Would you like an example of Leaping into new Meadows? This chapter reads more like mathematics than most of the book. Alas, it muddles along in a line, which is exactly the opposite of what I want to convey about the ephemeral and multidimensional nature of the Leaps in math. But how can I write except one word after another? It's not the fault of math, but of language!

It's okay if you don't explicitly understand all the math in this chapter. This is not primarily a math book. It's a book for parents about sharing the math they already know.

As long as the mathematical content of the Meadows makes sense, please keep reading. Quickly is okay. If the math gets "beyond you," skip to the conclusions on page 46 and then go on to the next chapter.

As one example of how math blooms, let us examine a concept that every reader knows—that of number. In particular, we will track the number three. Follow me to see how three changes as the learner Leaps across the Brooks of Mathland.

Each Meadow listed below requires new understanding beyond the preceding Meadow. Going from one to the next requires a Leap of understanding, similar to that experienced when learning to ride a bike. Each Leap needs its own fertile setting. Effective teachers provide settings that help children ripen mathematically and take the next Leap. Teachers cannot Leap for children, although they (and classmates) can point the way and provide encouragement. As can parents.

Each Meadow has its own secrets. The "practice" suggested for each Meadow indicates some of its treasures. Notice how the practice activities reinforce the understanding of that Meadow and provide insight that may lead to the next Leap, without pushing the practicer into a sinkhole. If the Meadow's philosophy is understood, a moderate amount of practice will be pleasant—interesting but not intimidating.

Do we have to wander through the Meadows in the order given? Not necessarily, but this is one good order. We certainly do *not* have to explore any Meadow completely to begin enjoying the next. We can go back later and explore hidden flowers if we want to bound over a large territory before examining details.

For methodical spirits, this chapter outlines one satisfactory program for becoming acquainted with many views of the number three. But restless wanderers can move on quickly, and even jump ahead or sideways. Learning math is not as linear as this chapter suggests.

Someone who misses any of these Meadows altogether will develop math anxiety before graduating from high school. Computations that don't make sense are usually the culprit. If a Meadow isn't your own, struggling with exercises that the locals do comfortably makes you feel like an alien. You don't know the local dialect and dance, and you can't seem to keep up. Anxiety grows where understanding belongs, and you want to run away.

Meadow 1: As a child is learning words and sounds, it is fun to say "1, 2, 3, 4, 5. . . ." At this stage it's merely a chant, like "pat a cake, pat a cake, baker's man . . ." Both are fun. Both are ways that adults share and laugh with young children. You and your child are running across the Meadow of unadorned words, delighting in their **sound and rhythm.**

Practice: As you do any repetitive activity with a child (such as taking steps, putting away toys, handing out snacks), count rhythmically.

Meadow 2: The words "1, 2, 3, 4 . . ." can be **attached to objects** that are in some sense together. The child sees a group of objects and "counts" them. Adults are pleased.

Everyone laughs and hugs. The count may not be accurate. The child may skip an item or two. Or count some of them twice. No matter. The child is putting number names to objects and, at the same time, is discovering that number names are not attached to individual things the way "chair" and "baby" and "tree" are. Number names can bring light to the eyes of grown-ups when applied to *any* object.

Practice: Enjoy your child's efforts to count alone and in concert. Encourage starting the process by saying, "Let's count these spoons. One, two . . ." Do not comment on inaccuracies. Just pay attention and enjoy the sharing.

Meadow 3: The child **names objects** with numbers, thereby counting them. There is just one number per item, and one item per word. The number "3" names only one of the items during any one counting. "Three" is not blurred over two of the items. Counting is distinct.[2]

Practice: Ask your child to count small sets of objects. "How many people are eating dinner together today?" If the child makes a mistake, you might suggest counting again. The child will probably be willing and may count correctly the second time. If not, don't worry about it.

Meadow 4: The child realizes that **whole sets of objects can be named** by a single number. That is, every time you count a set, you get the same answer no matter in what order you take the individual objects. Educators say that at this stage a child has learned **conservation of number.**

Practice: Ask your child how many things are in some set:

▸ How many chairs are in this room?

▸ How many carrots are on your plate?

▸ How many children are in this picture?

If your child counts incorrectly and you have time, you may want to ask for a recount. If not, no big deal. Accuracy is not your goal, just familiarity and fun with numbers and counting.

That Three-Letter Word

Some parents clutch at the word "set." Please don't. You know about sets of dishes and sets of tools. A particular set is whatever Daddy and Mommy say it is. If we count the set of dishes on a table, that's what we're counting—because somebody nearby decided it would be counted. Of course, children should be allowed to decide what sets are worth counting too.

The point is that every set has exactly one number associated with it: the number of objects in that set. For example, how many dishes are there in your family's *set* of dishes?

If your child counts correctly, fine. If your child says the correct number without counting, better yet! At this stage you want to develop "number sense," a feeling for numbers independent of counting. If you have time, you may want to ask how she knew.

More practice: Make cards with dots or stars on them in standard arrangements, and learn them by sight with your child. The arrangements on dice and dominoes are basic. Can your child recognize these patterns without counting?

You don't have to count to know that three is associated with triangles, waltzes, and tricycles. Soon your child won't either. Numbers have meaning apart from counting.

"Little children count. We look for patterns." I said this often to a first grade class. In April the remedial teacher told me one of the boys quoted my phrase to her. He explained that she should be looking for better ways of doing math problems than just counting. She and I were thrilled. I'm not against counting, but life is short and art is long.

Meadow 5: Numbers can be put together and always come out to the **same answer,** no matter what objects they are describing.

- One apple and two apples are three apples.
- One pencil and two pencils are three pencils.
- One bean and two beans are three beans.
- One apple and two oranges are three pieces of fruit.
- One tree and two bushes are three plants.

We write this $1 + 2 = 3$. At this stage your child doesn't know the "facts," but he does know what addition is. Numbers have become sufficiently abstract that adding two of them makes sense.

Practice: Now you can practice addition facts while driving around town. Your child will be eager to learn the rhythms like a grown-up. They make seeing patterns easier. You don't have to count so high if you perceive parts of a set and add them up. You are also less likely to make mistakes.

Subtraction follows quickly after addition with all of its interpretations. In some sense subtraction is addition backwards. Multiplication and division are not far behind. Children need to play with these processes before they memorize facts, but once they grasp addition, they can see that other processes might yield other facts. Subtraction, multiplication, and division are three new Meadows, but in this metaphor, they are alongside addition. We travel in a different direction:

Meadow 6: The child uses numbers to **measure**. Three inches can be seen as a one-inch length three times in the same direction.

|_____|_____|_____|

Practice: Measure things. Leave rulers and tape measures around where your children can measure things. How many inches wide is the kitchen table? How many feet? How long is the picture of a house that your child just drew?

Estimate lengths. Ask your child to make an estimate too. Then measure to see what the real length is.

Make scale drawings of rooms in your house. Use them to contemplate rearranging the furniture.

Meadow 7: The child "places" numbers on an abstract **number line** where the old familiar "1, 2, 3, 4 . . ." are like clothespins among which other numbers are hung. They are called "counting numbers."

$$0 \quad 1 \quad 2 \quad 3 \quad 4 \quad 5 \text{ ---} \longrightarrow$$

Practice: Help your child place "new" numbers like ½, ¾, and 1⅓ on the number line amidst the old friends. The newcomers are called "fractions" or "rational numbers."

("Rational numbers" is a nineteenth-century term apparently used to distinguish fractions from the other numbers that lie on the number line, which are called "irrational numbers." $\sqrt{2}$ and π are two of the best known irrational numbers. They can't be written as fractions.)

Be sure that zero is on the left of your number line and that the numbers extend (theoretically) endlessly to the right. This is only a convention, but it is an Earth-wide convention.

Meadow 8: Adding three to any other number is understood as moving three places *right* on the number line.

Practice: Interpret additions by arrows on the number line. For example, 2 + 3 can be interpreted as one arrow of length two and another of length three. Head to tail, they total an arrow of length five:

Meadow 9: Subtracting three from another number is understood as moving three places *left* on the line.

Practice: Do subtraction facts using the number line. What happens when you subtract a larger number from a smaller number?

Meadow 10: The counting numbers are the clothespins of only half the number line. The other half goes left from zero and contains **negative numbers.** Their applications include temperatures, holes in the sand, basements in a large building, and streets south of Main Street. Adults find money owed a useful interpretation of negative numbers.

$$\longleftarrow -4 \quad -3 \quad -2 \quad -1 \quad 0 \quad 1 \quad 2 \quad 3 \quad 4 \longrightarrow$$

Practice: Investigate adding, subtracting, multiplying, and dividing negative numbers.

Meadow 11: Three is **a motion of all numbers** three units to the right along this number line. It can be understood as a rigid motion of the line to the right. It can be written $y = x + 3$ or $f(x) = x + 3$.

Meadow 11A: Three is a motion taking all points **three times as far from zero** as they originally were. It can be understood as stretching the number line away from zero: the farther a point is from zero the farther it moves. It is written $y = 3x$ or $f(x) = 3x$.

Practice: Interpret other simple algebraic expressions and equations as motions on or of the number line.

In Meadows 12 and 12A relationships such as $y = x + 3$ or $y = 3x$ are plotted on a **graph in two dimensions called the x-y plane.** The horizontal axis is always x, and the positive numbers go right—just because everyone else does it that way. The vertical axis is the y-axis, and the positive numbers go up for the same reason. It's nice to have international unanimity on one issue. We humans all agree on this convention!

Meadow 12: The equation $y = x + 3$ is graphed as a line that is three units up from the line $y = x$, a line that is at 45° to the axes and goes through the origin (the point where the two axes cross).

Meadow 12A: The equation $y = 3x$ "is" a line through the origin with a slope of 3. That means that it is going up three times as fast as it is going right if you measure the vertical and horizontal distances corresponding to any two points.

Practice: Graph other linear equations (that is, equations whose graph is a straight line), watching their y-intercept (how far from the origin they cross the y-axis) and their slope (the "rise" divided by the "run" of any two points).

Meadow 13: The graph of $y = f(x) + 3$ is three units higher on the x-y plane than the graph of $y = f(x)$.

Practice: Interpret the relationship of various graphs to each other.

Conclusions

I could go on, but you get the idea—even if you don't understand every word. The concept of three—indeed the concept of numbers—keeps changing and expanding.

It's not that anything in the earlier concepts becomes false. Although we may be comfortable in many Meadows, we still enjoy the jingle "1, 2, 3, 4, 5 . . ." of Meadow 1. But motions and equations involving each of these numbers play increasingly significant roles as our mathematical sophistication grows. At each stage, some practice is needed. But at each of these stages, a new perception of truth must seize each person. We cannot piggyback on someone else's understanding. We must construct each insight for ourselves.

The same is true for mathematical concepts beyond numbers. They too change and grow, yet the original meanings do not become false. A similar chapter might be written about any other number or any other mathematical concept. I chose three in part because it is a familiar concept to parents, but also because I have been especially aware of my own growing understanding of three as I share math with children. A decade ago, I didn't see nearly as many complexities in three as I now see.

You may learn from this chapter that even simple math is harder than you thought. Even simple ideas are far more complex than they seem at first. What a challenge your child has!

You may know the first few Meadows as well as you know how to comb your hair. That's because you are no longer an alien in those Meadows. But when faced with lots of details, even you gulp. Remember to be kind to those who don't yet have the overview. Each aspect of math is hard when you're an alien. Even three.

C h a p t e r

Tending the Garden: Questions and Answers

5

Adults may think that medicine must be painful in order to be efficacious. Children are perfectly happy to enjoy mathematics. Adults may think that the good things of life are either illegal, immoral, or fattening. Children believe that thinking about interesting mathematical questions in a creative and intelligent way is fun.

The paid attendance at baseball games would not be increased if we required every fifth grader to memorize the current batting averages of every player in the Red Sox lineup. Nor would the understanding of the game be increased. The true flavor of mathematics, like baseball, is subtle and elusive.

—*Robert Davis*[1]

What can adults do to help children's mathematical lives bloom? If our children are flowers, we are the gardeners. We cannot force the flowers to bloom, but we can tend the garden so that blooming is much more likely. One of my hobbies is raising my own family's vegetables year-round without any poisons or artificial fertilizers. Sharing gardening ideas is fun. Sharing mathematical ideas is fun. Sharing ideas is great fun, but neither gardening tips nor mathematical ideas will guarantee blossoms. Flowering requires individuality from those using the ideas. What works for me may not yield blossoms for you. Take from my ideas whatever is useful, and figure out the rest for yourself. I have faith that you can find a good way for your family.

There is only one rule that is immutable, at least for young children: Both you and your child *must enjoy your math*. Stop before your child has had enough. If your child associates mathematics with conflict or anger, the big game is basically lost. Two aspects of this immutable rule are *listening carefully* to your child's explanations and *not worrying about mistakes*—neither your own nor your child's. Nobody's giving grades at home.

Providing a setting where children will have Wonderful Ideas is crucial. Since each child is different, there is no "right" way. There are several approaches, including toys, games, drawings, practice, and conversation. The particular toys, games, and drawings you provide will depend on the age and mathematical maturity of the learner. They will be discussed in age-specific chapters.

Conversation is vital at all ages, and the guidelines are surprisingly similar. Before your child can speak, your conversation must be a monologue. You should talk continually with your pre-verbal child, and some of the conversation might as well be about numbers and shapes.

However, after your youngster can talk back, long monologues are not well received. Even at the graduate level with highly motivated mature students, I try to keep my lectures from exceeding five minutes. People stop paying attention, and a lecturer can soon violate the cardinal rule of teaching: Keep the student wanting more.

Questions

When people ask me why my students learn so well, I tend to reply, "I ask questions. I try never to 'tell' preschool children or graduate students the answers." Questions! They are the key. This is not the whole truth. Often, I plan carefully what I am going to say and ask. Often, I collect or make manipulatives or diagrams. I am always seeking ways to convey ideas from my mind to the minds of others. Planning is central to good teaching. But the opening, the glue, and the closing are questions.

I fear being misunderstood. Questions play many different roles in our lives. The questions that teach math are not torture questions; they are pleasure questions.

Psychologist David Elkind warns that parents of MacArthur fellows[2] "appear not to have applied unreasonable pressure." Is asking questions unreasonable pressure? Is sharing your love of learning unreasonable pressure? Or is it encouraging a child's innate "rage to learn?"[3]

"No quizzing," admonishes Mary Leonhardt repeatedly as she advocates the "fun and easy" way of teaching your preschooler to read.[4] But is questioning always quizzing? I think not. Can quizzing be fun? I think so. If we ask a young child a quiz question, we have to be willing to answer it or drop it before the child becomes bored or frustrated. Only occasionally do we need a young child to answer any particular question. The preschool years are not a series of tests. The elementary grades shouldn't be either.

There are several different kinds of questions. We might distinguish informational questions, attention-getting questions, rhetorical questions, listening questions, quiz questions, and thought-provoking questions. Let us examine each in turn.

INFORMATIONAL QUESTIONS

Informational questions are used when you don't know the answer and want it. "Did Grandma call while I was out?" "Did you eat all your breakfast?" "Have you brushed your teeth?" "Where is your left shoe?" "Where is my pocketbook?" These questions are part of daily life. As our offspring grow older, the questions may become more sophisticated. Do you remember who was president in 1876? What was the name of that space probe that went out beyond Neptune?

Informational questions are the prototype questions, but only a small fraction of spoken questions are really seeking new information for the inquirer.

ATTENTION-GETTING QUESTIONS

We all remember without fondness the bores who stand up at the beginning of a question-and-answer period following a public program. They seem primarily to want attention but may also be using their questioning to share an idea. This is not information gathering. Sometimes callers to radio talk shows have similar non-informational motives.

Families often use attention-getting questions. Instead of a command, you might say, "Would you like me to help you with your coat?" This may direct your child's attention to the fact that you want to leave soon.

> I was at a gathering in a home and overheard a father ask his five-year-old, "What is the biggest number?"
>
> "Infinity," said the child.
>
> "Oo! oo!" cooed those nearby, impressed. I decided not to barge in and ask the child about infinity plus one. An impression was being made.

RHETORICAL QUESTIONS

"When are you going to clean up those toys?"

"Don't you know how angry it makes me when you scatter your shoes around the living room?"

"Do you remember what your mother and I told you about TV?"

"When am I ever going to get some peace and quiet around here?"

These too are not really questions, although they end with question marks. They are closely related to attention-getting questions, but they are shorter and often not premeditated. They are expressions of feelings or statements of ideas.

LISTENING QUESTIONS

"Why do you think that?" may be the most basic question in the teaching of math. Learn to ask it without giving a clue as to whether you agree or disagree with what was just said. You will learn amazing things from and about your child.

Listening questions are similar to informational questions, but the information you seek is temporary. People's thoughts and understandings change.

Listening questions prepare us for our next leading question. To lead someone, you must know "where" they are now. Often, the next leading question will be another listening question, but not always. Listening questions probe another person's mind. Their answers cannot be right or wrong.

"How did you get that?"

"Why do you think that?"

"Are you saying that . . . (restating what you think you heard in your own words)?"

Master mathematics professor Heinrich Brinkmann was known on the Swarthmore College campus for being able to find something right in what every student said. No matter how outrageous a student's contribution or question, he could respond, "Oh, I see what you're thinking. You're looking at it as if . . . but should you consider . . . ?" or "Maybe you thought we meant . . . but can you try . . . ?"

This takes real insight. It takes careful listening, and it also requires a repertoire of potential meanings. Affirming someone by articulating the accuracies in their confusion is one way to create a "teachable moment"—a time when students are especially receptive to learning.

If you listen carefully enough to find something right in everything your child says, you will find yourself learning mathematics along with your child. This is satisfying because there is always more to learn. Hearing your child's thoughts is a way to thoroughly digest the mathematics you already know and expand your horizons.

Remember that everything other people say has meaning to them. There *is* something right in what each person says. To guide your child, you must first recognize the blockages on the road to understanding. You must hear what this young person is experiencing—this person who is in some ways a continuation of you into the next generation but is also profoundly different.

A parent needs to listen with an open ear and a silent voice. Reflecting upon both facts and feelings is helpful. You learn so much! You want to find out what your child is thinking. Mathematically, a listening parent or teacher revisits old ideas in mysterious new clothing. Each child comes up with amazing insights and questions. I am always startled by how much first graders can teach me. They take little for granted. They have fresh glimpses into this subject that we can't see, hear, feel, taste, or smell.

If the child's answer is wrong, you could put that fact inside your head for future reference. You don't have to address it immediately, although if you have time, you may want to ask another probing question. Or you may want to remember to chat about it later. Don't criticize your child for wrong answers. A child should be safe at home—except for informational questions, where you really need the right answer.

Listen for both your child's sake and your own. You are both trying to achieve new levels of understanding and also to clear away cobwebs that cloud the cave of comprehension. What questions can you ask to brush away the cobwebs? Since all math leads to the same place (for all people in all cultures), understanding the cobwebs and misconceptions in your child's mind is important for clearing the path to the next Meadow.

What Is a Number? Once my son and I were walking down the street discussing topics like $2 + 0$, $3 + 0$, $2 - 0$, and $3 - 0$.

"Zero is an interesting number, isn't it, Ed?" I commented.

"Zero is not a number," said my companion firmly. This interested me. I had taught about infinity enough to know that in some contexts it seems like a number and in others it doesn't, so I recognized Ed might be having similar ambivalence toward zero. I wanted to explore his idea.

"What is zero, Ed?"

Without missing a beat, he replied, "Zero is the thing that keeps the numbers from getting mixed up with the minus numbers." [5]

"Glad I asked," I thought to myself! I had been tempted to simply insist that zero was a number. By asking him to explain his understanding, I learned that he knew more than I realized, and I also understood his view much better than I had before.

Listen especially to see if your child is using words the same way you are. An entire chapter on this follows. Definitions are often not helpful. You need to use words in context correctly and often for your child to pick up the meaning.

Many mathematical troubles occur because someone doesn't understand "the question." Sometimes this is because the listener isn't paying attention, but usually it's more subtle. The potential for misunderstanding in mathematics is awesome. If you can tap into the real thoughts of the person before you, you can untangle the knots around their mathematical inner light. If you can help your child be patient enough to dissect questions carefully, you have built significant mathematical ability (and also test-taking ability).

TEST OR QUIZ QUESTIONS

Test or quiz questions are controversial. Is it okay to quiz your child? Some people suggest it may be abusive. I think it depends on the context and spirit.

Everyone agrees you should not judge your own children. You are not going to expel them from your home or make the decision to place them in a special program. There is no need to be judgmental. Your own children are always cherished.

How children respond to quiz questions depends somewhat upon their age. Teenagers generally do not appreciate being quizzed. They get enough of that at school.

However, during a child's second year, many parents and children enjoy the "trained dog" routine. "Lie down and roll over!" Good child! Hugs and kisses. It's fun to see if the child understands the command enough to follow through directions.

Many parents of young children enjoy teaching their children nursery rhymes and songs. I can't see how it could be more damaging to also sometimes chant, "Mommy has two ears. One, two. How many ears does baby have? One, two." At a later stage the quiz might be, "Daddy has five fingers. One, two, three, four, five. Can you count my fingers?" If it works, fun! If not, it's fun to be together. It's a game.

The spirit is crucial. If you can't do it without getting anxious, it's better not to do it at all. Keep it light. Interaction and attention are fun. If the quizzing is lighthearted, your children will enjoy these "tests." If they associate "whoopee" when the answer is right and friendly attention when it is wrong, they develop a taste for quizzing and a sense of inevitable success. Surely that will be an advantage later.

One of the basics you want to teach is that **it is all right to be wrong.** Your child will still be loved. Failure is a part of life. Frustration is normal. Failure precedes success. It may lead to triumph. If your child retains that conviction that all toddlers have, you've set the stage for future mathematical success. A child who can't be comfortably wrong isn't going far.

There is absolutely no need for the parent of a preschooler to worry about whether a child's answer is right or wrong. If the answer is right, you can rejoice. If not, try again later. No big deal.

If you get upset with wrong answers, you should not ask your child quiz questions. But probably you can learn to play. You are introducing your child to a pervasive aspect of our culture.

Sometimes children enjoy quizzing grown-ups. This is fun too. Play along. Give your child the sense of power that comes from asking you. Some parents enjoy pretending they don't know and letting the child correct and teach them. Do what is fun for you.

Preschool quizzing is like playing house, where nothing can go seriously wrong. When I was a child, all little girls played house because they saw their mommies keeping house. Now preschoolers see folks going to school and taking tests. Playing at quizzes does not have the emotion of taking quizzes, any more than playing house has the emotions of actually keeping house. Playing at quizzes prepares children to be more relaxed with the real tests later—*if* it is fun.

THOUGHT-PROVOKING QUESTIONS

"In the United States, the purpose of a question is to get an answer. In Japan, teachers pose questions to stimulate thought. A Japanese teacher considers a question to be a poor one if it elicits an immediate answer, for this indicates that the students were not challenged to think."[6] Japanese teachers report that much of their time together is spent discussing which questions effectively stimulate thought and discussion.

Thought-stimulating questions are the jewels of good math teaching. "Incorrect solutions . . . become topics for discussion in Asian classrooms."[7] They can provoke playful discussion in American homes too.

"Why do you think that?" is a good response whenever your child makes a statement that you think is wrong. It may not seem so wrong after you hear your child's explanation. It may give you an opportunity to state what is right, affirm your child, and establish another teachable moment.

Thought questions, no matter how playfully presented, are more than play. They open up new avenues of inquiry. They stimulate thought.

"I wonder if . . ."

"I wonder what . . ."

"What do you think would happen if . . . ?"

"Can you tell me how you thought that up?"

"How did you get that?"

"Does that make sense?"

"Can you think of an easier problem?"

"What easier problem does this remind you of?"

"Can you explain it another way?"

"Can you think of another way?"

"Can you think of another way?" is a super-valuable thought question. There is always another way! Japanese teachers encourage children to suggest alternative ways to do a problem. Then they compare the methods and discuss which are the best ways before solving the problem. Some thought-provoking questions direct the conversation more specifically.

"How many plates would fit on this tray?"

"Are there more people in the park today than yesterday?"

"How many songs can we sing until Daddy comes home?"

"How can we arrange these flowers?"

"How could we find the number of carrots in this bag?"

"How many pretzels should each child get?"

"How many times do we fill this cup with sand to fill the bowl?"

"Which one of these containers contains more water?"

"Why do you think so?" Again and again: "Why do you think so?"

"Can you bring me three diapers?" If you ask a thought-provoking question that is also a request, be sure there is enough time to cope gently with an error. You don't want to upset the child or yourself. Easy does it. If there is plenty of time, you can reply, "Thank you. That is two diapers. One, two. Can you get me number three so I will have one, two, three diapers?"

Teaching questions also gain information—about what is going on inside the child's mind. As we listen to the answers to thought-provoking questions, we consider how to tweak the child's thinking. "You are really working hard on that!" we may say. Our appreciation of both the child and the problem's difficulty may be conveyed in our demeanor. Other good thought-provoking *comments* include:

"I don't understand, either. Let's look at it together."

"What a good idea!"

"What a good problem! How did you think of that?"

Often children enjoy setting math problems for themselves.

Giving Fruit: Gretchen Cuccio and her four-year-old nephew were handing out fruit to the people sitting around the table. Each time they gave a piece away, they counted to see how many they had left. The boy found this very exciting. "Five minus one is four!" he would exclaim.

"How many glasses are on this table?" could be either a quiz or thought-provoking question, depending on the stage of the child. A thought question for a young child becomes a quiz question later. The category depends on the child. There is a difference in spirit, however. An asker of a thought question does not need to know the answer. "How many people are in this room?" "How much will it cost to feed our Thanksgiving gathering?"

Sometimes when you think you are asking a quiz question, you discover it is really a thought-provoking question. Your child is missing something. Can you find the concept and address it? One of the challenges of parenthood is recognizing that a concept is difficult for your child and responding appropriately. This, of course, is true far beyond mathematics!

Questions focus mathematical attention. Overcoaching has become a cultural habit, but Socrates in ancient Greece had the right idea. Try not to give answers. Let your students discover them.

Right Answers?

If you let your child discover answers, they will not always be right (even when there is a "right" answer). While people are contemplating a new math concept, gathering courage to Leap another Brook, they must concentrate on the ideas and the process. It is your job to provide a setting—to be sure the garden is fertile, but not to judge.

Do your children have the space, time, and quiet to really concentrate? Can their thoughts travel along their own paths? Do their surroundings allow and help them to focus? If so, they will want to find right answers, and will often do so, but often they won't. What then?

Remember that partial learning is okay. There is no mathematical understanding that is vital today. That's good, because our knowledge is always incomplete. Homes should be accepting, so if the struggle does not yield the right answer today, we hope for tomorrow.

If the focus is on right answers, real learning may suffer. "Learning to argue about mathematical ideas is fundamental to understanding mathematics. Chinese and Japanese children begin learning these skills in the first grade; many American elementary school students are never exposed to them." [8]

Posing one's own problems is an essential part of becoming competent and confident at math. Encourage a child to ask, ask, ask. Children are naturally curious and will keep asking questions if praised for doing so.

You won't always answer the questions if you are otherwise occupied. Just murmur an appreciative response like, "What an interesting question!" If you are busy or need time to think about the question, try, "Ask me later when I have time to give you the attention you deserve." If the child remembers to ask again, you have a budding mathematics buff. If not, try to remember yourself.

You and your preschooler may enjoy arguing about mathematical ideas. It's a pleasant kind of arguing, not to be confused with conflict. "Conversation" may be a better word to describe it, but it does include batting ideas back and forth like a tennis ball. When our schools neglect this activity, it is all the more important for families to supplement. There are times to put right answers in abeyance and enjoy a kaleidoscope of ideas.

Of course, in the long run, we all care very much about right answers. The issue in this section is how to prepare children to solve complicated problems correctly when they are older. It is not by insisting upon right answers today. It is by expecting them to explain how they obtained their answers. Children enjoy being "right," and they will struggle for right answers for the right reason—the pleasure they bring. We don't have to push.

In the short run, whether an answer needs to be correct depends on what type of question we've asked:

▸ When we ask for **information,** we want right answers. If you didn't brush your teeth, you must do so before jumping into bed.

▸ **Attention-grabbing** and **rhetorical** questions don't need *any* answers. When we ask, "When are you going to clean up your room?" nonverbal room-cleaning is better than any verbal answer.

▸ **Listening** questions evoke right answers by definition. There is no wrong answer to "Why do you think that?"

▸ **Thought-provoking** questions do not need an *objectively* right answer. Their purpose is to stimulate creative thought.

▸ **Quiz** questions eventually need right answers, but not during preschool, and maybe never at home. Home quiz questions should always be just play. When a child gives a wrong answer, either the child is bored or the question was too hard. In the first case, it is time to move on. In the second, it is time for more discussion or games.

Estimating answers, even for information and quiz questions, is becoming more important as problems become more complex. To "estimate" is to find a number near the answer, but not necessarily exactly right. Estimating is especially needed to check calculators. For example, if you buy one item for 29 cents and another for 42 cents, it is valuable to know the total is *about* 70 (= 30 + 40) cents. This is a better estimate than 90 cents or 10 cents—and much better than 71 dollars! Helping children estimate is important—often more important than seeking right answers.[9]

Of course, if you are home schooling and you expect that someday your offspring will attend a school or college, you eventually want to be sure they can produce right answers on demand. But stick to a reasonable calendar. Don't push. Pushing can become a counterproductive habit. Your child came to you wanting to learn, and that "rage to learn" will continue if you tend the garden cheerfully.

Eventually, little children become aware of outside standards. They usually become more concerned with being right. People like being a respected member of society. Children will try to get right answers unless they become convinced it isn't "cool."

Typically, students become more anxious about tests as the stakes grow higher. If, however, they are confident, they may not. One very gifted child was diagnosed to lack "the anxiety he'd need to compete with the other gifted kids."[10] It is sad that some adult gatekeepers believe that people must be anxious to achieve. I believe a little anxiety goes a long way.

Adolescents may need the visible outside demands that seem necessary for adult social and economic competitiveness. Certainly, we all know that unless your adolescent gets many right answers on school tests and national exams, prestigious colleges are out of reach. That may be important to you. Children, however, have an internal urge to achieve until it is socialized out of them.

Weeds? Sleep Deprivation, Sugar, Allergies, and TV

Like every other garden, the garden in which we raise children has its weeds, although we may question what are "weeds" and what are misplaced "flowers." Obviously, children don't learn math as well as they might if they don't have adequate sleep, exercise, or good food. Less obviously, we all thrive best mathematically with human math companionship; we were meant to learn in groups.

For many people, eating sugar decreases the ability to concentrate. The effects of meat, salt, fat, herbicides, pesticides, bovine growth hormones, food coloring, and preservatives are controversial. It isn't easy for Americans to eat health-building food, but it's worth a try.

Allergies increase the ante. I knew an otherwise agreeable child who became vicious when he ate wheat. His mother had a terrible time persuading schools not to feed him wheat, even when they telephoned to complain about his behavior after doing so. Individual variations are apparently unbelievable to some folk, despite repeated evidence.

TV has a terrible reputation in academic circles, but some of its offerings are excellent. It seems to be here to stay. Parents should try to monitor what their young children watch. News can be pretty traumatic, maybe too much so for preschoolers. Time spent watching TV is obviously not spent reading, conversing, exercising, playing games, or doing puzzles or homework.

Surveys by the Educational Testing Service indicate that the typical U.S. sixteen-year-old has spent more time watching TV than attending school.[11] This report emphasizes that North Dakota has the highest scores on the NAEP (National Assessment of Educational Progress) Math Proficiency tests and the lowest percentage of its eighth graders who report watching over six hours of TV per day. Then it observes that North Carolina has almost the reverse: the highest percentage of eighth graders watching over six hours of TV per day in the continental states and the second lowest scores on the same tests.[12]

Does this indicate that abstinence from TV causes higher math achievement? First, we can wonder if the two tests actually measure what their designers intended them to measure. Perhaps North Carolina youngsters are more willing to *admit* their fondness for

TV. And what parents really know how much TV they (or their children?!) watched yesterday? Last week? (Stevenson and Stigler report that Japanese children watch more TV than U.S. children, but their surveys yielded very different hours per day from those found by ETS. They claim that Japanese parents are more likely, however, to require that homework be finished before watching TV.[13])

Assuming the surveys and tests are accurate, an applied math problem then arises. If high amounts of TV-watching and low math ability really do occur together, it may not be that one causes the other. Perhaps something else causes both—like lead poisoning, or parental indifference, or boring schools, or a community that disdains intellectual pursuits. More advanced math is needed to explore these questions effectively.

What Are the Causes? Cause and Correlation

How do we determine what causes what? Human beings, including little children, are constantly wondering *why* things happen. I believe that playing games cultivates math power and that too much TV damages it. Can I be sure?

Mathematics provides a way to measure how often some event accompanies another. This is called *correlation*—how often two events occur together. Correlation is a sufficiently complicated concept that a course in statistics is necessary to know the mathematical basics. (Such a course could be offered in middle schools if the teachers were prepared.) However, the rough idea involves computing a number between -1 and 1 to measure how often two events occur together. Two events that occur randomly with respect to each other are said to have "correlation zero." If they tend to occur together, they have positive correlation. If they seem to avoid each other, they have negative correlation.

If one event often follows another, we sometimes conclude the first *caused* the second. But that's not always so. For example, children's shoe sizes and grades on math tests have a positive correlation. The larger the shoe size, the higher the grades more often than not, but it's the children's older age that causes both. Putting little children in larger shoes won't cause them to learn more math. Teaching them math won't make them parade around in larger shoes. Neither causes the other.

Similarly, as I mentioned in Chapter 1, families with higher socioeconomic status tend to have children who do well in mathematics, but this does not imply that socioeconomic status causes math excellence. Family behavior patterns that promote math excellence currently correlate with socioeconomic status, but they can occur in any family. It is possible to change this correlation.

For another example, "men" and "engineer" currently have a positive correlation. This means that they tend to occur together. Most engineers are men, and more men than women are engineers. Conversely, "women" and "engineering majors" have a negative correlation in this country. However, "women" and "engineer" will have a high correlation at a meeting of the Society of Women Engineers (founded by Lillian Gilbreth, mother of the Montclair, New Jersey family hilariously depicted in *Cheaper by the Dozen* and the only engineer on a U.S. stamp). "Men" and "engineer" need not have a high correlation. Neither causes the other.

Since many people who smoke eventually get lung cancer and many people who get lung cancer previously smoked, many people conclude that smoking causes lung cancer. The tobacco companies claim that only a correlation has been proved *mathematically*. That is true. Autopsies of former smokers' lungs and cell research give more convincing proof of causation than math could possibly provide.

Those without a math background too often confuse correlation with causation. If two events (like TV-watching and low math grades) tend to occur together, one *may* cause the other. Then, again, maybe not. They may be linked by some common factor, or the juxtaposition may be coincidental. This is important for even little children to know. Things that happen together do not necessarily cause one another.

Patience with Your Child and Yourself

Developing patience is basic. Be kind to your child. Be patient with your child.

Normally, I am not a patient person, but when doing or teaching mathematics, I must be patient. It always startles me when someone comments on my patience in my professional capacity. Patience is the absolute bedrock of teaching mathematics.

Miracles don't occur on human schedules. When we teach mathematics, we are trying to evoke understanding, and it may be only a few seconds away. We don't want to jeopardize its arrival. It is amazing how much can happen mathematically if a teacher cultivates a feeling of leisure.

> I learned this "skill" at the feet of the aforementioned Heinrich Brinkmann. Dr. Brinkmann, a busy department chair, had the reputation of having an endless amount of time for any student's question. It wasn't so. It just seemed so. His courses were difficult, and his students learned rapidly, but he cultivated the *feeling* that there was nothing he would rather do than respond to your question. This feeling seemed to pry open the doors of understanding.

Patience with your child is paramount. In the preschool years, there is absolutely no hurry. Later, keep asking yourself what the hurry is. Thinking up things for oneself is fun. Don't deprive your child of that.

How can we develop patience? If you too are not a basically patient person, how do you wait patiently for youngsters to learn math? For me, the device is remembering that this is one of those moments life is for. Savor it.

Savoring mathematical moments is part of the play we all enjoy with little children. What *is* life for? What could be more important than waiting for your own child's intellectual growth?

Yes, yes. Getting out of a burning building and getting to work on time and . . . Still, the inner dialogue is worth a try: "I am a parent. My child is just a child. What a privilege! Life is here and now. It does not matter whether my child learns quickly. Speed does not matter. What matters is that we share this moment. That my child continues to tolerate frustration. That my child accepts failure, and knows that love is patient and kind."

Alas, this self-sermonizing doesn't always work. You can't always be patient with your child, sometimes because of outside demands and sometimes because you just don't have it in you. Then you need to be patient with yourself. There is just so much you can cope with.

Before you push your own limit, tell your child you must move on. It isn't the child's fault, but you can't pay attention to important things right now. The immediate demands of life are upon you.

Be kind to yourself. Be patient with yourself.

GETTING STARTED WITH YOUR CHILD

Fun and Games with Preschoolers 6

In games, feedback comes from other children and oneself. . . . In games, children check each other's thinking and learn that they can figure things out for themselves.

Children become mentally more active when there is the possibility of outdoing their opponents, or of being outdone by them. . . . In games . . . immediate feedback comes directly from their friends.

—Constance Kamii[1]

An ounce of motivation is worth a pound of skills anytime.

—David Elkind[2]

The cardinal rule is that both parent and child must approach math joyously, as the superb game that it is.

—Glenn Doman[3]

Should we teach our preschoolers math? Two of the most outspoken proponents of the "nay" (David Elkind) and "yea" (Glenn Doman) answers, respectively, are quoted above. Their quintessentially American agreement overwhelms their disagreement. If you want to explore their argument, their books are readable and interesting, but I will confine myself to their common messages. Whereas some cultures associate early math with misery, American writers of all stripes advocate and/or expect pleasure. Disagreements exist about how much little children can learn and how joyfully. Parents must listen attentively to each child to hear what that child wants to learn today. My own dissensions are with (1) the pessimism of the naysayers and (2) the specific procedures of the yea-sayers. These are not trivial quibbles.

However, our common agreement that early childhood should be the Age of Innocence outweighs any disagreement. If we all pulled together where we agree, our culture would take a great leap forward. I believe that it is possible (and not difficult):

- with great joy and love
- with little parental time
- to teach most children enough mathematics before kindergarten so that they shine by current standards.

I know this is possible because I have witnessed many parents with a mathematical mindset sharing basic principles of mathematics with little children. I believe our approach can be used by less "mathematical" parents. I have great faith in parental imagination and caring. If you listen to your own child, contemplate your own goals and daily life, and seize teachable moments, you can give your child a major head start without expending much time or money. This chapter and the next describe how.

Why I Advocate Early Childhood Mathematics

Obviously, I am biased on behalf of early learning.

The negative reason: Part of my reason is cynical: the U.S. school system is so unreliable, we must give our children every possible advantage to endure it. Having been a teacher myself, I concluded before giving birth that many otherwise fine U.S. teachers will do serious harm to children's *mathematical* capability if their families don't make them "school proof." Glenn Doman uses the phrase "school proof" to warn parents that unless they actively cultivate their preschool children's love of learning, common school practices may destroy their innate ability and their "rage to learn."[4] My interviews of black mathematicians corroborated this conviction.[5]

My sympathies are with Mary Leonhardt, a high school English teacher in one of our country's wealthiest suburban districts, who came to a similar conclusion *after* she had failed to teach two of her own children to read. She thought a love of reading would be enough, but found that because her son didn't learn to read immediately when he began first grade, he was placed in a "slow" reading group and developed self-esteem problems that seriously interfered with his learning. Later she was able to use her own professional skills outside school, and he become a college honors student, but she now urges parents to teach their children to read before school so they avoid the Leonhardts' needless suffering.[6]

Similarly, children with some math power when they begin school please the kindergarten teacher. They are labeled "smart," at least informally, and have advantages that are not bestowed on less fortunate youngsters. This happens in almost all U.S. schools.

The positive reason: Math is fun! Little children *want* to learn. If we slip a math question in from time to time, they will work hard to solve it. They like puzzles and problems. Not until someone teaches them that work is unpleasant do they resist learning and helping.

> In supermarket lines sometimes a preschooler near me gets cranky. Counting the number of people in line can be quieting. Sometimes I put up two fingers of one hand and one finger of the other and count the total number of fingers with the child. They like the attention, and usually stop whimpering. The incredulous parent watches. Adult Americans think it odd to play math games in the supermarket line, but preschoolers (and most children in the primary grades too) find it entertaining.

Many such experiences have convinced me that eagerness to learn math does not depend on genes or social class. Supermarket lines are inclusive. The statistically significant differences in five-year-olds' achievement measured by international studies are *not,* I am convinced, due to race or genes. That average U.S. five-year-olds are inferior mathematically to their counterparts elsewhere can't be due to genes. Collectively, ours are the same as those on other continents. But our culture neglects the joy of math.

All little children find math fun. Math is a great vehicle for giving attention and making people happy. It's cheap too.

Theories abound about when and what to teach your children, but there seem to be two pieces of unanimous advice. All American "experts" advise parents of preschoolers to:

1. Talk with your kids.
2. Have fun with them.

How do parents have fun with their children? One way to interact is by playing games. Traditional pastimes that brought so much fun in yesteryear surely can't hurt much. For those who weren't raised on these traditions, I describe below the games that I played as a young child. You can find others in stores and catalogs, and can make up similar games if you have more time than money. If you have time, these games are both pleasant and educational. Otherwise, the next chapter suggests ways to give your child math power with little, if any, "extra" time.

Games

If you know these games in other forms, that is fine. All versions have good mathematical content, and variety may make them more fun. Feel free to adapt and change the games to suit your family. Enjoy!

DOMINOES

Dominoes are small rectangles made up of two squares. Each square has dots on it. Usually there are between zero and six dots on each square. Larger sets have up to nine dots.

Put the dominoes face down on a table. Then each player picks seven dominoes, and puts them on their sides so only the player who selected them can see them. The first player puts one of his dominoes face up on the table. The next player puts out one of hers, matching a domino square to one already on the table with the same number of dots. This continues with each player setting out a domino in turn.

At first, each player has two choices, one on each end of the train of dominoes. However, if a person plays a domino with the same number of dots in both its squares, that domino is played at right angles to the train, centered at the end of the train. This gives three new places to play—continuing the original train and at each sideways direction. At first all three directions offer the same number, but as new dominoes are put down, more numbers appear. Thus, the number of options typically grows as the number of dominoes in front of each player diminishes.

If a player does not have a matching domino, that player takes one from the face-down dominoes until a play is possible. The first player to get rid of all dominoes wins.

The game of Dominoes teaches number recognition and matching. The numbers appear in a pattern that each player eventually recognizes instantly. This recognition develops an understanding of number (Meadow 4) that is beyond counting.

BOARD GAMES

Simple board games, such as the classic Chutes and Ladders, are games in which players roll dice and move markers the number of spaces shown on the dice. The boards have special instructions on some of the spaces, so the child learns the concepts of "go backward two spaces" and "move one more space forward." The player whose marker gets to the end first wins.

These games, which you can either buy or easily make, teach addition facts up through six, the concept of counting forward, and the concept of subtraction as motion backward. It is not easy to tell where the concept of counting ends and addition begins. Simple board games teach both. They prepare a child for Meadow 6.

CARD GAMES

War is a card game played with a bridge or poker deck. It can be a partial deck after some of the cards have been lost or damaged. Two or more can play.

The dealer gives one card to each player in turn, around and around the circle, until all the cards are in the hands of the players. All players make a deck of their cards, and place the deck face down.

For each turn, all players turn over the top card of their deck. Everyone looks at all the cards, and the person who has the highest card takes all the facing cards and tucks them under her or his deck.

If the highest cards in any one play are the same, the people who played these cards turn over their next card, and the card that is highest gets to "take" all the cards on the table with faces up.

When someone runs out of cards, the person with the most cards wins.

The game of War teaches "larger than" and "largest." It is totally a game of luck, so your child will win just as often as you do if you play together. I loved this game as a preschooler. I can't imagine why.

Peace is the same game as war except that the smallest number on a card wins the hand. It teaches "less than" and "smallest."

Concentration: a deck of cards is placed face down on a table. You can use a much smaller number of cards, perhaps beginning with only a few pairs. Play rotates around the table.

For each play, one player turns over two cards while the other players see what they are and where they belong. If the cards are the same number (or two jacks, queens, kings, or aces), the player keeps them. Otherwise, the player puts them back exactly where they were. If a player makes a match, another try follows immediately. Otherwise, the play moves on. The game ends when the players have taken all the cards, and the player with the most cards wins.

This game teaches numbers recognition and, of course, memory and placement. It is a good game with homemade dot cards that teach sophisticated number recognition. For example, a card with six nines will be picked up with one that has nine sixes.

Go Fish: Each player is dealt six cards. Matching pairs of cards are put down in front of each player, and are replaced from the deck. Play rotates around the table.

Each player addresses another specific player and asks if he or she has a particular number (or jack, king, queen, or ace). If so, that player must hand over the card. If not, the response is, "Go fish." The player picks another card from the deck. If it is the one asked for, another such play occurs. Otherwise, the next player plays.

As cards are matched, they are put down in front of each player. The game ends when the cards are all used up. The player with the most cards (in pairs) wins.

This game teaches number recognition and memory. It also involves luck and psychology, so the winner is not as consistent as in Concentration. A common adaption is to wait until all four cards are in one hand before they are put down on the table.

Anything a child learns before kindergarten is gravy, so there is no need to feel pressured or goal-oriented. Enjoy. Never push for mastery. Don't jeopardize a goose that lays golden eggs—joy. The preschool years are a time for play. Our job is to help children retain their inborn desire to learn and their faith that they can.

Nurturing Preschool Promise

Virtually all young children like mathematics. They do mathematics naturally, discovering patterns and making conjectures based on observation. Natural curiosity is a powerful teacher, especially for mathematics.

—Everybody Counts [1]

The most important pattern for preschoolers is that math is fun. Nothing is worth violating this principle! Little children know this basic truth, and parents should never allow math to add tension to their preschooler's life. There are no specific goals. Goals begin in kindergarten. If either you or your child isn't in the mood, skip it.

You want to preserve your child's innate tolerance of frustration. Together you want to explore ideas, wondering about numbers, shapes, and other patterns. You want to talk math once in a while, but only voluntarily. How can it be harmful to talk about math, if you are both having fun?

Teaching math to your preschooler need not take much (if any) extra time. With the help of this chapter, those on "Tending the Garden" and "Language," a little thought, and (preferably) some conversation with congenial adults, you can teach math to the child beside you as you work. You can integrate math into daily activities. It's a matter of recognizing and using teachable moments.

Integrating Math into Everyday Activities

Every time is the right time to do math. On laps. While walking. While setting the table. While cooking. While shopping. Math cheers up waiting rooms and chauffeuring.

On laps: As a child tugs at your ears while cuddling, you can say, "Daddy has two ears." Tugging appropriately, "One ear, two ears!"

While walking: "Four more houses until we're home. Now it's three more houses until we're home. Two more houses. One more."

I fell in step with a two-year-old and her grandmother while crossing the street after waiting for the red light together. "One, two, three, four, five, six . . . ," I began counting our steps and soon grandma was chanting too. The two-year-old beamed with the rhythm and pretended that she too was counting.

While setting the table: "There are four plates. I put down one plate. Now I have three plates. I put down one plate. Now I have two plates. . . ."

While cooking: Count the potatoes while scrubbing them. Or: "I put a pot on the stove. I put another pot on the stove. Now there are one, two pots on the stove!"

While shopping: "We need three loaves of bread. One, two, three." Or: "One, two, three, four cans" as you drop them into the shopping cart. Or: "The cereal is in aisle 7. We are at aisle 3. We have four more aisles to go."

In waiting rooms: "How many blue chairs in this waiting room? How many red chairs? How many chairs altogether? How many pictures on the wall?"

The Hand Game: If you have four pencils or pennies, you can enjoy the time you spend in waiting rooms with children—yours or others. Show everyone the total number of objects you have. Put them all behind your back, and then bring some back into view. Let the child guess how many are still behind your back. Of course, you can play it with three, five, six, or any other number of items.[2]

In the car: Jane Gaertner was driving with her four-year-old and asked how her imaginary dog was doing today. "Fine. She had 10 puppies, 7 girls and 5 boys." Jane considered how to respond, when Kate piped up again. "Oh, no, 5 and 5 make 10. She had 5 girls and 5 boys."

Fun Drill

All little children enjoy "fun drill." Yes, indeed, drill can be fun! Fun drill is very different from Drill and Kill. It is drill that entertains. Many families enjoy pointing to things and having little ones say their names, one form of fun drill.

Some possibilities of pleasant math drill, assuming the children are at an appropriate stage, are:

▶ Counting the stairs as the toddler "walks" up them with the help of a caretaker.

▶ Counting the apples in a different order to see if they count to the same number either way.

- Counting the number of people in a checkout line. Every time one person comes or goes, count again.
- Dividing the pretzels, grapes, raisins, or other bedtime snack among three siblings each evening.

Each of these activities might be considered "Practice." If done with joy, they build the understanding that prepares a child for another intellectual Leap. Let us examine each:

- The child knows she is going upstairs and thinks that words are fun. The drill teaches a special set of words—numbers.
- The child counting apples knows the names of the numbers, but hasn't realized that each group of objects corresponds to exactly one of these numbers. Educators call this teaching "conservation of number." (The number is "conserved" whatever order you count the objects.)
- Repeatedly counting the number of people in a line teaches the relationship between counting, adding one, and subtracting one.
- The children dividing the bedtime snack realize that there is some "fair" number that each child should have that evening—that each day's problem has a unique answer. As they divide the given number by three each evening, they learn their three times table. They are also learning the concepts of division and multiplication.

In each case, the child is repeating a familiar idea, but it is not boring. The child knows the basic concept and is learning more about it. The drill is life-giving. Such drills are not Drill and Kill *because they are fun and meaningful.* Healthy drill is valuable and can be pleasant.

Geometry and Visual Patterns

Traditionally, math has been envisioned as two interwoven ladders: arithmetic-algebra and geometry. Knowing one helps you learn the other. One way to teach both is to collect used bottles with their matching caps. Challenge your child to put each cap on the bottle it fits. This teaches one-to-one correspondence, eye-and-hand coordination, and geometry. The child also has to consider how many pairs are left, so subtraction is also subtly being learned.

Let's consider the geometry strand first.

During your child's second year, try to acquire some simple puzzles. Geometric shapes that fit through holes and make a noise as they drop are simple and fun. "Doughnuts" of graduated sizes that can be stacked on a common stem teach visual understanding of "greater than" and "less than."

Blocks are absolutely essential for both boys and girls. "Blocks are a fundamental learning material for young children," says writer-psychologist David Elkind.[3] Indeed, the difference between male and female "abilities" in mathematics may well be due to the difference in their playing with blocks and other construction toys. (As a tot, I was encouraged to play with my brother's.)

Modeling clay, Play-Doh, and sandboxes provide creative geometric exploration. Every child should have access to these; every family can afford mud, if nothing better. You can make inexpensive Play-Doh by wetting equal quantities of flour and salt. Children can play happily for hours while they learn about the structures of our three-dimensional world.

These toys are more important for your child than anything else that you can buy except books. Pattern Blocks, a new-fashioned toy, are also exceptionally valuable. They are brightly colored triangles, squares, trapezoids, rhombuses, and hexagons that fit together in many ways.

Browse through the catalogs listed in Appendix C. Their contents include wonderful presents for birthdays and holidays. Toys designed to teach children math are called "manipulatives." The big name allows administrators to buy them for schools. In schools, Pattern Blocks are manipulatives. For birthdays, they are toys.

HEXAGON!

As my brother's son, Adam, approached his second birthday, I sent him a set of Pattern Blocks for a birthday present. His mother sat down with him that evening and taught him the names of the shapes of these colorful blocks. When my father arrived to celebrate his birthday a few days later, Adam proudly showed him the blocks, calling them by their proper names.

Soon Adam was identifying shapes in the natural environment. My brother telephoned to tell me that he was surprised at how many hexagons (six-sided figures) there are in real life. His best story was about the local post office floor—a goldmine of hexagons.

"Hexagon! Hexagon! Hexagon!" Adam jumped from one glorious discovery to another. His shrill two-year-old voice greatly amused the postal patrons who thought they had come for prosaic transactions. What better messenger of mathematics than an enthusiastic two-year-old?

Before long, Adam shouted "Hexagon!" at a stop sign. His parents looked thoughtfully at the sign, and informed him it was an "octagon" (an eight-sided figure). A rueful father told me a few weeks later that Adam could now distinguish between hexagons and octagons faster than his parents. When they saw such shapes, they had to either count or think, while Adam identified them correctly by immediate visual recognition.

> A hexagon is a six-sided figure and an octagon is an eight-sided figure. A "regular" figure has all sides and angles equal. Pattern Blocks include regular hexagons. A stop sign is a regular octagon.
>

There are cheaper ways than Pattern Blocks to teach about shapes. Parents with experience and some comfort with mathematics can take bolder risks.

SANDWICHES AND SHAPES

When Dr. Susan Epstein made sandwiches for her preschool children, she would first ask them what kind of jelly they wanted with their peanut butter. Then she would ask them what shapes they wanted.

"Three triangles and a square!" they might say. Then she cut the sandwiches into the requested shapes. The children enjoyed trying to stump her. Could they make a request Mother couldn't fulfill? She liked developing her ability to cut sandwiches into the requested shapes without any leftovers.

A less mathematically confident adult might share puzzling through problems that their children posed, using cooperation instead of competition. Letting the children pose the problems gives them a sense of power and creativity while providing practice with shapes. If you need more control, you might want to suggest the shapes yourself and then help your child figure out a way to do it. What is important is that both adults and children enjoy the process.

Be creative in finding ways to include geometry in your interaction with your children. Sewing, "cutting" cookies, doing home repairs, and planning simple carpentry projects all provide abundant opportunities to chatter about shapes. Arranging dolls' houses and toy car parking lots give other possibilities. As long as you keep it light, it doesn't matter whether it's age-appropriate. Visual patterns are entertaining. If you keep babbling about the geometry in what you are doing, occasionally your child will notice.

Pre-arithmetic

Recognizing numerals and rote counting, "one, two, three, four . . . ," is not really math, but knowing these helps children explore mathematics faster after they have learned the *concept* of number.

COUNTING ALOUD

Apparently, teaching children to count before they understand numbers has been around a long time. Two *Mother Goose* rhymes are clearly designed for this purpose:

- ▸ 1, 2, 3, 4, 5, I caught a fish alive. 6, 7, 8, 9, 10, I let him go again.
- ▸ 1, 2, buckle my shoe. 3, 4, shut the door. 5, 6, pick up sticks. 7, 8, lay them straight. 9, 10, a big fat hen.

Both rhymes teach rote counting to ten. Notice also that the first rhyme teaches two fives are ten and the second teaches five twos are ten. $5 \times 2 = 2 \times 5 = 10$. The commutative law of multiplication in *Mother Goose*!

Your child can learn numbers higher than ten while walking and climbing. During the second year, walking up stairs is a laborious but satisfying achievement. While the child concentrates on muscular strain, the accompanying adult can be counting each step aloud. I remember vividly that there were 14 steps between the first and second floors in the home where my little brother and sister grew up and 15 in the home where I raised my own children.

One young father told me that when they want to change their two-year-old's diaper, they say, "Let's count the stairs!" His child then runs to the bottom of the stairs, ready to climb to the diaper-changing site on the second floor. He doesn't care about soiled diapers, but counting stairs is a treat.

How young can one begin such activities? I associate them with the second and third years, but why worry?

I watched Bev Clark count, "One, two, three" as she put tablespoons full of formula into her seven-month-old's bottle. Wendy stopped crying when Mother said "one," watched with interest, and showed signs of visible relief when the third spoonful had been deposited. Then there was only the screwing of the nipple onto the bottle until it was plopped into the eager but silent mouth. We can't say to what extent Wendy knew that three tablespoons were necessary to relieve her hunger, but this baby did seem to know that "one, two, three" had pleasant connotations.

Counting anything that repeats, such as strokes of a washcloth or buildings that we pass while driving down the street, entertains the parent (at least) and familiarizes the child with words that will be useful later. A more sophisticated form of this exercise is asking school-aged children to count some designated item during long trips (cars with Illinois license plates, pedestrians wearing shorts, dogs, gas stations, Corvettes, etc.).

The Pollak Game entertained mathematician Henry Pollak's two children for years. One child was posted on each side of the car. They got 10 points for each horse or cat that they saw, 5 points for each cow, and 1 point for each chicken. Each dog sighted lost that player 10 points. Other animals were worth specified numbers of points. The first child to reach 100 points won.

RECOGNIZING NUMERALS

Another pre-arithmetic achievement that might begin in the second or third year is recognizing numerals. Numerals can be seen. They represent numbers, which can't be seen, heard, felt, tasted, or smelled. The difference between numerals and numbers resembles the difference between teaching your child the words of the catechism and expecting the child to understand the meaning. Numerals and words are much easier to learn than the meaning behind them.

If your blocks have numerals and letters on them, mention the names as you play with the blocks. Plastic refrigerator magnets teach children to name letters and numerals. You might point to numerals and letters in books as you read to your child.

Karen Bernard taught eighteen-month-old Rachel to recognize numerals and letters on her blocks. One day, when Karen opened the refrigerator, Rachel excitedly pointed to the milk carton and screamed, "M!"

As children begin to recognize and name letters and numbers (not necessarily consistently), it is entertaining to point out numerals in other contexts—on the fronts of houses, in supermarkets, on signs displaying the temperature or time, and labeling floors in elevators. If parents read numerals out loud, eventually children join the fun. Like rote counting, this is not really mathematics, but it makes math easier when the time comes.

By the time Axel was two, he enjoyed entertaining shoppers by shouting out the letters and numerals on supermarket signs.

I have no idea how developmentally typical Adam, Rachel, and Axel were, and it doesn't matter. These are skills your child will be taught in kindergarten, so there is no hurry. If your child learns them at age four, she or he will still be deemed precocious in kindergarten. However, if you play "Naming Shapes," "Counting," and "Naming Numerals" with your one-year-old, she or he may pick them up along with general language development. Why not?

Understanding Numbers

Real math begins with the concept "two." When your child commands, "Two crackers!" with the obvious expectation that you will hand over more than one, you have a budding mathematician. It's an exciting moment. How do we set the stage?

Each time a caretaker picks up two shoes or gloves and has nothing better to say, "One shoe, two shoes!" or "One glove, two gloves!" is good conversational prattle. The chatter that loving adults heap on children can include the counting of ears, eyes, arms, hands, feet, and legs.

The Magic of Two: "How many eyes does Kim have?"
Kim parrots, "One, two," pointing to each eye in turn.
"How many eyes does Mommy have?
Kim points to Mommy's eyes. "One, two!"

This is primarily for fun, but it may help Kim develop the concept of two. Eventually we reach the stage where you ask, "Do you want one cracker or two?" The first few (or many) times, the child babbles nonsense. Cheerfully, you hand the child one cracker.

Finally the child mimics, probably at random, "Two!" The parent hands the child two crackers. Not immediately, but eventually, the child realizes that saying "two crackers" has a payoff that is worth the effort. When this happens, the child has begun to learn real mathematics.

When the child makes the connection between "two crackers" and "two carrots," your child is using patterns to solve problems.

Months later, probably, the child realizes that "three crackers" is an even better way of manipulating the environment than "two crackers." Soon it is time for really exciting ideas that any adult recognizes as math.

Now is the time for counting that is not rote.

Axel's father turned over the truck that Axel was thoroughly enjoying. "How many wheels, Axel?"

Axel sat back in anticipation of his feat. "One, two, three, four!" he said triumphantly, pointing to the wheels in order. He grinned at Daddy and me, and we grinned back.

Aggie Azzolino, author of *Math Games for Adult and Child*[4] and mother of Michael, suggests that once a child has a rudimentary knowledge of number, asking for two spoons or five spoons (or three paper clips or four pencils) is a good game to play when the parent is grading papers, emptying the dishwasher, or watching TV. This is a life-giving drill. Michael knew what numbers were and wanted to practice.

This is the kind of practice—*following* the understanding—that helps mathematical development. The child knows the answer much of the time and wants to confirm his or her knowledge. That's fine. Such drill is healthy for all concerned. Mommy may find it a bit boring, so Aggie Azzolino suggests you play many of the games suggested in her book while doing something else. Obviously, you shouldn't play unless your child finds it satisfying.

It is often more than a year after children can count four objects until they can reliably count ten or twenty objects.

Addition

When your child can count accurately small sets of items, you can begin addition. My daughter remembers loving this activity. We did it frequently in sessions of less than a minute. Sometimes we did it for more than a minute on lines and in waiting rooms.

TEACHING THE MEANING OF ADDITION

I would hold up a few fingers in each hand. "How many fingers?" I would ask. I might hold up two fingers on my left hand and one on my right.

My daughter would count the total number of fingers, and I would say, "Good! That's right! Two plus one equals three." While I said, "two," I would wiggle the two fingers on the left hand. While I said "one," I would wiggle the one finger on my right hand. While I said "three," I would wiggle all three fingers.

You can play this game for all addition facts up to five plus five. It can be played while waiting for the washing machine to finish its cycle, and while waiting for another child to join the carpool. It doesn't take any equipment except a parent's fingers. It's always handy. Once a child has learned to count to ten reliably, it guarantees the child consistent success in the game "Quiz." *After* the child provides the answer, the parent verbalizes the problem. "Good! That's right! Five plus three equals eight!"

What can be the harm? It takes no time, except time when you're waiting around. It's fun. The child associates math with success and learns easy addition facts quickly.

There is only one caveat: don't start playing this game until your child can count accurately up to the total number of fingers that you hold up on both hands. (In other words, you can start before the child can count accurately to ten, but if she routinely skips six, don't use more than five fingers on both hands.)

TEACHING ADDITION FACTS

After a while, your child beats you to the punch line. He will say the fact himself. You hold up your fingers, and your child will say, "Three plus two equals five." Your child has reached Meadow 5, and you can begin to teach addition facts without props.

By this I mean that your child understands the small numbers and the concept "plus." He or she knows the word "equals." When you say, "What does six plus eight equal?" it makes sense. Your child doesn't know the answer as she or he enters the addition Meadow but knows how to figure it out. The child may not get the right answer, but recognizes there is one right answer. When you say, "What makes you think that?" there will be some response.

Now you can play car games. As you wait for red lights and drive down easy streets, spout out questions like, "What does seven plus four equal?" The follow-up question is "How did you get that?" whether the answer is right or wrong. Your child is learning to "talk mathematics."

My second child was only nineteen months younger than my firstborn, so we soon had a competitive game. (Competition is not necessary, nor maybe even desirable.) We gradually moved on to subtraction, multiplication, division facts, and simple algebraic equations. Everybody seemed to enjoy it. I don't think fun math drill does young children any harm *as long as everyone is having a good time.* You can play these games while waiting in lines, while carrying in the groceries, and in other odd moments. I recommend them especially while driving because they keep the kids happy. As I have said before, I don't advocate any particular timetable.

Logic and Transfer of Knowledge

Some math is neither geometry nor arithmetic. Logic relates to connections between abstract patterns—learning a pattern in one context and seeing its power in another. Preschoolers can begin to comprehend abstractions.

My Daughter and the Dryer's Dots: One day as I was taking the laundry out of the dryer, my preschool firstborn began counting the holes in the dryer's door. While I drew item after item out of the dryer, she counted the holes, on and on.

I watched, impressed at the curiosity of the young, until it was time to take the clean laundry upstairs. Then I suggested that she count only the number of rows and the number of holes in each row. Since each row had the same number of dots, we could multiply and find the total numbers of dots without counting so high. I told her that I knew the answer, and she could learn it. Then she could easily find the number of holes in any dryer's door whenever she wanted to.

My daughter still remembers walking up the stairs behind me with a heady new sense of independence. Math was not just a game where you got the right answer to please Mommy! It was real, something she could do by herself to find out things *she* wanted to know.

She looks back at that moment as a time when she gained a whole new mindset. She now observes that lots of kids get stuck in the mindset that math is something done to please parents or teachers. They "see math as just dots or oranges, or whatever their parents or teachers use." Seeing math as "real," something you can use for yourself, is pivotal.

Developing logic, decision-making capability, and relationships to applications can begin before a child remembers. Stay alert to possibilities. Each child and each family is different. How can you teach about logic, decisions, and applications? Here are two examples of children too young to remember the excitement, but who clearly felt the sense of accomplishment that accompanies solving a challenging problem.

Which Dessert? When my daughter was thirteen months old, I had two different desserts available. I held them in front of her, one in either hand.

"You may eat one of these," I said. "I will eat the other. Which one do you want?" She looked from one to the other. It was the first conscious choice of her life with permission from an adult, and she understood the responsibility of the moment.

After a while, she reached for one.

"Okay," I said, pulling the other toward my mouth. "I'll eat this one." She looked at it longingly and reached. I offered both hands again.

"Okay, you can have this one. But then I get that one." I retrieved the first. Of course, soon she wanted it back.

We did a few cycles of this. Both the baby and I were considering each step thoughtfully without any evident anguish, but it was too much for my husband.

"You're the parent," he chided. "Why don't you just decide which one she gets?"

"Because some day she has to learn to make decisions and live by the results, and this time I know she won't be hurt by whichever decision she makes." That made sense to him. He too watched patiently while we went back and forth from one to the other. The quest was not about which dessert she ate, but the process of choosing.

Eventually, she made a decision and seemed firm. Quickly, I ate the other dessert, and we were all happy.

The Flatware Problem: Suzanne Granstrand had heard that if you want your children to be helpful when they're eight, you must let them be "helpful" when they're two. Thus she allowed her two-year-old son to help empty the dishwasher by putting the silver away. Her knives, forks, tablespoons, and teaspoons have holes in the handles so they can be hung on a little stand. There are four hooks on each side of the stand.

At first Luke was challenged by merely slipping each piece onto some hook. His mother soon realized that it was a challenging math problem for a two-year-old to find the proper hook. After explaining his job, she busied herself with other activities, pointing out from time to time that spoons belonged together here and forks there. Gradually, Luke discerned the differences among the various types of flatware and placed them correctly.

After some diligent experimentation and his mother's tactful coaching, Luke could not only neatly categorize the pieces but also place all the spoons so they fit together with their bowls in the same direction. This took him several months, but he not only developed his pattern-perception abilities and his hand-to-eye coordination, but also got a feeling of accomplishment and the satisfaction of genuinely helping. Commercial puzzles teach the same skills. But after her son learned this one, Mother Granstrand had one fewer task to fill up her busy day. Her two-year-old could do a grown-up's job.

Music

Recent studies indicate that learning music often makes children better at math. Nobody knows why. Perhaps it is the exploration of abstract patterns, or learning to concentrate, or allowing oneself to "become lost" in a satisfying activity. Whatever the reason, if you enjoy music yourself, sharing it with your child may have the double benefit of helping your child become better at math. Many mathematicians are also good musicians, but some are not.

Flexibility

Use your imagination. Whenever you are with your child and not concentrating on something else, try to weave math into whatever else you are doing. Don't expect your child to understand everything you say—about math or anything else. Your child needs to hear your voice. Ideas float. Enjoy whimsy together.

Suppose I do it wrong? Suppose it's the wrong time? Then your child won't learn. As long as there is no emotional tension, no harm is done. Your child hears words all the time that don't make sense. By listening, children learn. Advanced mathematics won't harm your child as long as both of you are having fun. If your child is enjoying it, don't be afraid to overshoot.

The image of math as a ladder has been overdone. Sometimes people learn things out of order. I knew a baby who walked before he could crawl. I knew another child who could read before he could speak in full sentences.

Although math itself is somewhat hierarchical, the learning of math is not a straight line. It meanders around the Meadows. It keeps revisiting old territories with new eyes. If you and your child want to discuss something that's too hard for one or both of you, don't worry. Just keep it light. As long as you are both having fun, it doesn't matter how much sticks this time.

Your child may seem to forget everything, even if he or she once seemed to understand. Not to worry. You have accomplished three things. You've taught your child that math is a companionable subject. You and your child have realized how very smart she or he is. And buried memories will revive more easily later.

Don't underestimate your child. Don't expect him or her to learn on schedule, either. Preschoolers are avid learners, but unpredictable.

Language, the Slippery Bridge

8

Ten divided in two is five.
Ten divided in half is five.
Therefore, does two equal half?

— *A source of student confusion*
observed by Eleanor Wilson Orr[1]

Mathematics is more than fun. Sometimes answers are required—right answers at that. Right answers require that we grasp the *intended* meaning of the test questions. If we don't understand what is asked, we get "wrong" answers. Test scores plunge.

Eleanor Wilson Orr was one of the first to publicize the tremendous effect language has on learning mathematics. A private secondary school principal who won a grant to bring large numbers of urban children into her suburban school, she noted that she could integrate the two groups satisfactorily in obviously culture-based subjects such as language and social studies. However, bringing the newcomers up to speed in mathematics seemed, at first, virtually impossible.

Everything changed when she realized the extent to which language was interfering with her new students' participation in mathematics class. The title of her book *Twice as Less* comes from an expression that she often heard—one I too have heard since she brought it to my attention. What is "twice as less"? Two less than? Half as much? The original quantity minus twice its amount? Once Eleanor Orr focused on vocabulary and basic grammatical constructions, she could help her new students move forward mathematically. I found that applying Orr's observations yielded rapid mathematical improvement in urban primary grade classes.

If children learn basic vocabulary at home, they are at a great advantage in pleasing teachers and test-makers. Definitions are essentially useless for beginners. Young children must learn words in context, as we all must unless the new words can be defined using words we already know. The most basic words cannot be defined because we don't yet know even more basic ideas.

Words are slippery. Their meanings slip and slide, so when we put them together into sentences, we can't be sure that the listener will hear what we intended. Nevertheless, the more words we know, the more likely we are to communicate accurately. Especially in mathematics, if we don't know the conventional mathematical meaning of words, we can be needlessly left out.

When I flit back and forth from urban to suburban schools, the differences in children's mathematical vocabularies is mind-boggling. It has nothing to do with how fast the children learn; that appears similar in both locations. However, when children don't receive what I mean by crucial words such as "between," "each," and "same," my ability to share what I know is greatly curtailed. How much is their ability eclipsed?

Linguistic Causes of Mathematical Confusion

Language affects more than communication. It also enables us to toss around ideas in our own minds. Language facilitates conversations with oneself in our own musings.

Our vocabulary also affects how we understand problems that are posed to us. Alert teachers of mathematics at all levels are well aware that many—if not most—apparently mathematical problems are really communication problems. Misunderstanding words causes wrong answers.

DOUBLE NEGATIVES

"It's not that I never did that," implies "I did that," in "standard" English. Alas, some English dialects use "I didn't never do that" to emphasize "I absolutely did *not* do that." Subtraction becomes addition by a quirk of language.

Some languages, including Spanish, use the double negative in standard usage for emphasis. Native speakers of such languages may be at a serious disadvantage studying negative numbers in an English-speaking school. Traditionally, U.S. teaching of negative numbers assumes familiarity with the standard English double negative.

It need not be so. Opposites of opposites is another way to explain double negatives. Preceding a number with a minus sign is visualized as selecting the number on the other side of zero that is the same distance from zero on the number line. In other words, multiplying by -1 is seen as flipping to the opposite number with zero as the middle point. Then to multiply by any negative number, just multiply by the positive "part" of the number (the "absolute value") and take its opposite. Using this approach, a negative times a negative *looks like* a positive because flipping over zero twice brings you back to where you were.[2]

$$-4 \quad -3 \quad -2 \quad -1 \quad 0 \quad 1 \quad 2 \quad 3 \quad 4$$

$$\longleftarrow ------ \text{ opposites } ------ \longrightarrow$$

NOW AND LATER

Another linguistic pitfall for Spanish immigrants is that of "*ahora.*" In the Carribean islands *ahora* means "now," but in South America *ahora* means "later." This distinction can have serious consequences in word problems such as, "We are leaving now to drive 50 miles an hour for three hours. How far from home are we *ahora*?" Any parent or teacher of bilingual children needs to be sensitive to the possibility of linguistic treachery.[3]

Monolinguals have word hazards too. Chapter 4 traced how the meaning of a number expands without losing its previous meanings. "Sesame Street" teaches little children math words such as "above," "below," "inside," and "outside." In Chapter 10, we will explore the slippery meanings of "minus," "area," and "fraction." We now turn to some other words that befuddle many monolinguals.

BETWEEN

"Between" was one of the most important mathematical concepts that many youngsters did not understand, according to *Twice as Less*. It is a slippery word, but basic. How can we teach it? In a large class, I carefully choose three children to stand in front of the class. I would arrange them as shown in the cartoon on page 85.

Question: Who is between the other two in position?

Answer: Bernice, because she is right of Alex and left of Carla. Other words for "position" might be "location" or "place."

Question: Who is between the other two in height?

Answer: Alex, because he is shorter than Carla and taller than Bernice.

Question: Who is between the other two in weight?

Answer: Carla, because she is heavier than Bernice but lighter than Alex.

The word "between" has (at least!) three different meanings in this context, so the answer will depend upon which meaning is used. This "between" lesson takes about a half hour. However, after I teach "between" in third grade, teachers note rapid improvement

WHO IS BETWEEN THE OTHER TWO?

ALL ANSWERS ARE CORRECT.

SEE THE TEXT.

Alex   Bernice   Carla

not only in math but also in geography and phonics. How can children sound out words until they recognize when one sound is between others?

Another between lesson for a smaller group or a single student involves assembling three pens.

The felt-tip pen is between the other two in width because it is fatter than the ballpoint and thinner than the magic marker.

The magic marker is between the other two in length because it is shorter than the ball-point and longer than the felt-tip.

The ballpoint is between the other two in color because it is lighter than the felt-tip and darker than the magic marker.

4″ black felt-tip pen   5″ white, fat magic marker   6″ gray, skinny ballpoint pen

Of course, there is also another "between" depending on the location of the pens. Indeed, this can change before the children's eyes by rearranging the pens.

Young children will learn the *concept* of between (although perhaps not the word) if they have plenty of play involving "seriation," an educators' word for putting things in a series. Bottle caps and buttons are especially suitable for seriation play. They can be ordered according to diameter, height, or color. It is easy to collect an array of bottle caps in almost any U.S. household.

Music also provides opportunities to discuss "between" in pitch, in length of notes, and in order of notes.

Be alert for teachable moments about "between." If you are waiting in line with two children, you can ask them who is between the others. Use the word consciously as you speak. Think of different ways to use the concept, and discuss them with your children and with other housemates in your children's presence.

EACH

"Each" is much harder to teach than "between." "Between" is easy to remediate, once you realize its importance. However, when a child responds to the request, "Put two hats on each bird" by putting two hats on one bird, or one hat on two birds, or two hats on two birds, I have serious trouble teaching that "each bird" implies moving on to the third and fourth birds too.

If anyone finds a foolproof way to teach "each," please let me know. Meanwhile, parents can keep an eye out while using the word "each" in daily conversation to see if their children understand. None of my next suggestions, collected over the years, have been as uniformly successful as the above lessons on "between."

1. "Suppose I have one cookie in each hand. How many cookies are there in all?" Or, "I have one rubber band on each finger. How many rubber bands altogether?" These also teach one-to-one correspondence.

2. Write the word "each" next to "one." Contrast them in sentences: "One child is wearing a hat." "Each child is wearing a hat." Demonstrate the meaning of each (!) sentence.

3. In a small group of children, say, "Give one child a cracker." Then say, "Now give each child a cracker."

4. With drawings or blocks, try "Give each of your birds a partner." Then say, "Give each of your birds two partners." Teachers report that the word "partner" helps in teaching "each."

5. "Give one of your birds a hat." Then, "Give each of your birds a hat." Then, "Give each of your birds two wings." (This leads gently into "How many wings in all?" and the concept of multiplication.)

Remember the Brinkmann principle (see p. 50) of finding something right in each thing your child says or does. "Oh, I see what you are doing. You gave two hats to one bird. Can you give two hats to the next bird, and the next? I hope you can give two hats to *each* bird."

Children who don't understand "each" can't solve word problems such as, "If a teacher wants to give each of her twenty-five students three pieces of paper, how much paper does she need?" or "If each child is to get two sandwiches, each of which uses two pieces of bread, how many pieces of bread are needed for three children?" The arithmetic is easy, but the *math* requires understanding the word "each."

MANY AND MUCH

Some students confuse "many" and "much." "Many" is used when we are counting; mathematicians say the math is "discrete." "Much" can be used to measure length, or what mathematicians call a "continuous" variable. Examples are better than explanations:

How many pieces of paper?	or	How much paper?
How many pieces of bread?	or	How much bread?
How many yards of fabric?	or	How much fabric?
How many glasses of milk?	or	How much milk?

The distinction between "many" and "much" may be less essential than the meaning of "between," "each," and "same," but my experience suggests that people who misuse "many" and "much" also have trouble with the underlying mathematics.

LESS, FEWER, AND MORE

Children have more trouble with "less" and "fewer" than they do with "more." Perhaps it is because there are two words instead of one. Perhaps it's because "more" appears more in advertising. The old card game War, described on page 67, teaches "more." The suggested card game Peace teaches "less."

To understand the difference between "less" and "fewer," try:

Is there less water or milk?	or	Are there fewer ounces of water or milk?
Is there less silk or rayon?	or	Are there fewer yards of silk or rayon?

Is there less carpet or linoleum? or Are there fewer square yards of carpet or linoleum?

Is there less compost or lime? or Are there fewer bags of compost or lime?

In any of these questions, it would be correct to use "more" in place of either "less" or "fewer." In mathematical language, "less" modifies continuous quantities and "fewer" modifies discrete quantities.

SQUARES AND RECTANGLES

The following two figures both have four sides and four right angles. Therefore, the international mathematics community regards them both as rectangles. Everyone agrees that the top one is a square.

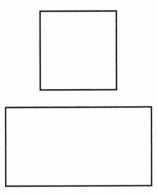

Is a square also a rectangle? Some books say, "No." These books claim that a square is not a rectangle. In my professional world, every square is a rectangle.

Language acquires meaning by convention. Words mean what we agree that they mean. Unfortunately, people writing math texts haven't *agreed* whether to consider a square a rectangle or not.

What do I tell my children? "Be wary! Language is slippery!"

Same and Different

Perhaps the most basic mathematical word is "same."

"But the very hairs on your head are numbered" (Matthew 10:30). "There is nothing new under the sun." These two common but contradictory sayings underlie a major challenge in mathematics. No two things are exactly the same, but everything is similar to something else. *Patterns* could be described as *perceptions* of sameness among difference. Mathematics is the study and use of patterns. Getting high grades on math tests involves figuring out what others perceive as the same, and using language similarly.

This is tricky. Anyone who has walked down a suburban sidewalk with a toddler must be amazed at the interest the youngster finds in every twig and pebble on the sidewalk. Surely toddlers see differences among twigs and pebbles that no longer fascinate the rest of us. If we say that two pebbles are the same, a toddler might, therefore, conclude that the word "same" means "different."

- ▶ Is the door of your room the same shape as this page?
- ▶ Are the points of a Christmas star the same as the tips of the fingers on one of your hands?
- ▶ Is the color of a clear summer sky the same as that of the ocean?

If they are described by the same word, does that make them the same? The slipperiness of the word "same" became especially obvious to me toward the end of my first year working in Newark, New Jersey.

After I had taught twenty weekly classes in a Newark first grade, the teacher and/or I spent ten minutes alone with each child, asking the child questions and listening to their answers. The children loved the individual attention and were surprised at the end when the teacher told them the sessions had been a test and they had done very well.

The second to last problem was to tell the total number of dots on a card like the following. Gold dots are symbolized by "g" and red dots by "r." The children had never seen a card with so many dots, although they had seen cards with 15, 20, and 25 dots grouped in fives.

```
g   g        r   r        g   g        r   r
  g            r            g            r
g   g        r   r        g   g        r   r

r   r        g   g        r   r        g   g
  r            g            r            g
r   r        g   g        r   r        g   g
```

To my delight, most of the children were able to say "forty" in May of first grade without counting. When I asked them how they had done so, most told me they had counted by fives in their head: 5, 10, 15, 20, 25, 30, 35, 40. Thus they knew, without my seeing how, that the answer was forty.

About a third of the children told me they knew five and five were ten, so they counted by tens: 10, 20, 30, 40.

I was very pleased, almost smug. These were children who did not have a good grasp of the concept of three when I had first visited the class the previous October. Now they were looking for patterns in a way that led to new concepts efficiently.

Pride goeth before a fall. I then asked the children how many dots there were on a card like the following. None knew.

```
g   g        r   r        g   g        r   r
  g            r            g            r
g   g        r   r        g   g        r   r
               r   r

r   r        g   g        r   r        g   g
  r            g            r            g
r   r        g   g        r   r        g   g
```

"What is the difference between the two cards?" I asked.

"They both come in fives."

"That's the same in both cards. What is different?"

"They both have red."

"That's the *same*. What is *different*?"

"They both have gold."

"That's the same. What is different?"

Only two of the children finally said that there were two extra dots in the middle of the second card, although the others must have seen them there, too. The two cards were side by side. The children simply didn't comprehend the meaning of the word "different!" I couldn't figure out how to teach them quickly.

(The two children who said that there were two new dots in the middle were not able to use this information to say there were 42 dots. The concept of starting at 40 and adding two was still beyond them. The teacher and I quickly remedied this once we realized the children needed to be taught.)

The problems with "same" are more than linguistic. Consider the corners of a door and the tips of the fingers on your left hand (excluding your thumb). They are the same number—four. However, the door looks very different from your hand. Sympathize with your poor child when you say they are the same!

Indeed, when your child perceives the sameness of two eyes, two ears, and two cookies, it is no mean feat. "Number sense" needs cultivation. In what sense are all of the following patterns the "same"? If you know immediately, that is because you already have "number sense." [4]

```
X X X      X X        X        X X        X X        X X X X
X X X    X     X     X X      X X        X          X       X
         X X       X X X          X X    X X X
```

TEACHING SAME AND DIFFERENT

We start by making statements using the words we want to teach. When we think the child understands the meaning of "same" and "different," we can ask, "How are these two toys the same? How are they different?" Be sure to use only teaching questions or play quiz questions, in the vocabulary of the chapter on "Tending the Garden." Try to enjoy the repetition yourself. People don't learn immediately. That's why children have parents.

> How are those two doors the same? They are both brown. They are both rectangles.
> How are those two doors different? One has a window in it and the other doesn't. One goes to the hall and the other to the closet.

Mathematics is full of opportunities to explore "same" and "different" with your child.

The two rectangles below are the same because they have the same shape. They are different because they have a different size. The top one has twice the perimeter (distance around the edge) as the bottom one. The top one has four times the area of the bottom one. (see pp. 117–19 for a discussion of "area"). It will take four of the smaller rectangles to cover the bigger one, but each side of the bigger rectangle is only twice the length of the corresponding side of the smaller one, so its perimeter is only twice the perimeter of the smaller rectangle.

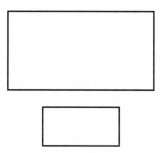

The two rectangles below are the same because they have the same area. (If you carefully split the top one horizontally and put the pieces side by side, you will get the bottom one.) They are different because they have different shapes and different perimeters.

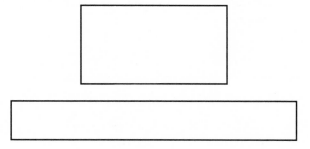

Some people may legitimately object to my calling these two rectangles the "same." It depends on the meaning of "same"—which is different in different contexts! For example, are the following two rectangles the same?

Yes, because they have the same shape, area, and perimeter.

No, because they are in different places!

Can anything be the same as anything else? Isn't everything the same as only itself? Things in different places are different in position! Pity the poor children trying to figure out the meaning of basic mathematical words.[5]

SAME OR DIFFERENT? FRACTIONS, DECIMALS, AND PERCENTAGES

Often when teaching either children or adults, I am asked whether a "different" answer is also right, and I respond by saying, "I see those two answers as the same. You are merely saying the same idea in a different way. Let's see if I can get you to see them as the same." Usually a student asking such a question is happy to accept my saying two similar things are "the same."

Unfortunately, I often worry that the student has made me an authority figure who can determine what is "right," when actually all human language is arbitrary. In these discussions, I would rather be seen as either a teacher of language or as umpire of the classroom "game," not as a purveyor of Ultimate Mathematical Truth. It is not easy to divest oneself of authority, even when you try.

When someone asks me whether it is "all right" to write ½ "in decimals," by which they mean .5 or .50 (or possibly 0.5000), I tell them that I see "½" and ".5" as two ways of writing the same number. Why should I care which language they use as long as they have the right answer? To me the following expression says that the numbers are interchangeable.

$$\tfrac{1}{2} = .5 = \tfrac{3}{6} = \tfrac{12}{24} = 50\% = \tfrac{50}{100} = 500{,}000/1{,}000{,}000 = \tfrac{1}{4} + \tfrac{1}{4} = \tfrac{2}{3} - \tfrac{1}{6}$$

They all represent the same ultimate meaning and so can be viewed as different languages for the same concept. Nevertheless, tests sometimes include commands such as, "Change ³⁄₆ to a percent." In my view, every number is already a percent. What such a command means is, "Write the number expressed by the fraction ³⁄₆ as a percent." It is unfair to be too harsh on the test writers. They are using common language. In a multicultural society, this inevitably has sad effects on those who don't share the subculture of the test writer, especially in multiple choice tests, where language is stripped of context.

Even commands such as "Compute ²⁄₃ − ¹⁄₆" are hazardous for the uninitiated. They assume that the reader knows that the preferred name for a certain point is ½ or .5. What "Compute ²⁄₃ − ¹⁄₆" really means is, "Find the preferred name for ²⁄₃ − ¹⁄₆." If you don't realize that ½ is just another name for the point on the number line designated by ²⁄₃ − ¹⁄₆, you may conclude that the hocus-pocus used to transform one expression to the other is just an annoying ritual made up by schools to distract children from real life and growing up.

Flexibility and Language

While it is crucial to teach our children language patterns, a parent needs to be flexible linguistically because:

- ► There are many ways to express an idea; language is subjective.
- ► Other people become alienated, even hostile, when we insist that ours is "the right" way to speak English. If you want your child to win friends and influence people, it helps to be flexible about language.
- ► To do well under the current system of standardized testing, a youngster must become adept at guessing what others mean.

Obviously, I use and recommend "standard" mathematical language most of the time, but at times I wish our culture acknowledged more widely the intrinsically imprecise nature of language. We should acknowledge that English is inadequate for communicating many important concepts and stop ridiculing leaders who try to express complex ideas but do so imperfectly.

How we learn depends on how we think. How we solve problems depends on what we understand them to be. Language is crucial to both. Furthermore, worldly success in mathematics (as measured by good grades, awards, and admission to prestigious schools) requires using the language of the current authority (parent, teacher, or test maker). School-age mathematics students try to figure out what the authority sees as the same in different contexts. This means knowing standard vocabulary and its variations. *No matter how much we know, if we can't communicate it, our knowledge is not much good to us and is no help to the rest of society.*

Language and Love

Words determine what a culture talks about. The abundance of mathematical words in English supports detailed mathematical discussions. Although I enjoy both the technology spawned by mathematics and math itself, I am uneasy about the biases of our language. English serves mathematics better than love.

For example, the Greek language has three words for our "love:" *filos, eros,* and *agape,* which, roughly translated, mean deep friendship, sexual love, and concern for all humankind. Tamil, the language of Southern India, has words for these three concepts and two more that translate into the English word "love:" the love between God and humans, and the love between a mother and her child. Some other Asian languages have even more words that translate into "love."

Thus, although English may pose problems mathematically, it interprets mathematics better than love. As you teach your child mathematical language in English, you are teaching the dominant language of the technological world. It is not so with love.

Specific Mathematical Words to Notice

It's important to learn some words as soon as possible including: two, plus, equals, zero, each, between, same, different, many, much, more, less, fewer, fraction, decimal, square, rectangle, circle, triangle, hexagon, right angle, parallel, diagonal, inches, feet, yards, ounces, cups, pints, quarts, gallons, centimeter, meter, area, and volume.[6]

Some of the words your child should learn in elementary school include: ten's place, hundred's place, digit, rounding a number, sum, product, quotient, place value, remainder, multiples, factors, common (as in common factor and common multiple), lowest terms, equivalent fractions, improper fractions, mixed numbers, coordinate system, angle, plane, acute angle, obtuse angle, perpendicular, polygons, quadrilaterals, pi, radius, radii, diameter, circumference, similar, congruent, perimeter, tons, miles, kilometer, millimeter, liter, and acre.[7]

The culture of mathematics uses some common words very differently than they are used in ordinary English. These present special pitfalls for the unwarned. Such words include odd, even, product, reduce, proper, improper, legs, mean, face, square, cube, power, degree, rational, irrational, real, imaginary, natural, law, differentiate, and integrate. An especially slippery word is "any," as exemplified by the sentence, "Is any even number prime?"[8]

YOUR CHILD STARTS SCHOOL

Primary-Grade Success 9

Time is a teacher's most precious asset.

—*Jay Mathews* [1]

Time is something your child's teacher doesn't have enough of. You can help by supplementing math time at home.

You have freedoms that your child's teacher doesn't have. You can pick the subjects you like and stop whenever either you or your child wants to stop—and keep going when both of you want to. There are fewer bells to interrupt your concentration or prod you to "work" when you aren't in the mood.

"Can I do it?" is the natural question of a math-anxious parent. I believe so.

How can I believe that every parent can help children mathematically when I am so adamant that teachers should have math preparation? Most important, parents are not setting curriculum. They are seizing teachable moments as they happen, responding to their children with active listening, and exploring patterns that have caught their own fancy. This is fine at home, but schools need more consistent expectations.

Second, parents have fewer children. To respond appropriately to dozens of children in the same room, a teacher must have many ideas at her mental "fingertips." You can afford to hesitate and explore.

Third, a parent can justifiably "learn while doing." If you consider yourself to be especially anxious when doing math, read one of the fine books about math anxiety. You don't want to pass anxiety onto your child, and you will benefit even more than your child. [2]

Your primary job is preserving your child's native enthusiasm and persistence. You are the cheering section, or at least a cheerleader. If you can also integrate math into daily life, that's even better. Listening, talking, playing, and asking thought-provoking questions are contributions you can make. [3]

Easy or Difficult?

"It's easy. *I* can do it." Mathematics education innovator Paul Goldenberg has noticed he has to remind his own children repeatedly that their being able to solve a problem doesn't mean it was easy. We all need to remind ourselves and our children.[4]

Parents of children beginning school tend to underestimate both how much mathematics they themselves know and how difficult it is. It's that old problem: after you know math, it seems easy. That doesn't mean it's easy for beginners! Your child is learning very abstract, difficult ideas.

If you haven't already read the chapter "Nurturing Preschool Promise," you may want to do so now. The school math curriculum *begins* with kindergarten. Your school-age child may enjoy many of the activities mentioned in that chapter if they aren't already too familiar.

Graduate Students and First Grade: In 1988–89, I taught three first grade classes each Wednesday morning, elementary calculus to undergraduates each Wednesday afternoon, and abstract algebra to graduate students, most of whom were high school calculus teachers, each Wednesday evening.

The first time I said to my abstract algebra students, "This morning when we were discussing this concept in first grade, we . . . ," everyone laughed heartily. The next week, we just giggled. As the weeks went by, we merely smiled as I transferred the concepts, conversations, and activities from first grade to graduate school. After a while, it seemed routine.

After the final exam, one student said, "I think we learned a lot more this year because you were also teaching first grade!" The others nodded emphatically.

The big surprise came when I graded the exams. The lowest grade was the same as the second highest the previous time I taught the course! Teachers don't make tests easier as they learn the material better, and neither class seemed more gifted. My graduate students had a better course because I was teaching first grade. The approaches I learned to reach the first graders were useful in higher math.

If you listen to questions and engage in respectful conversation with your K–4 child, you will be amazed at how much you too will learn. You will probably also be surprised at how much you can teach—and how much your child wants you to teach. Enjoy! These peaks of parenthood may not last forever.

As we noted in Chapter 4, the number three is not a triangle, a tricycle, or a pretzel, but these objects all have something that shows threeness. Furthermore, a "3" is not really a three, but just a symbol used to represent three. (Some math teachers think this is too difficult for nonmathematicians, but give it a whirl.) We might say that three is an "undefined term." Little children must learn many undefined terms with only a few clues about their meaning. What a job! We should not underestimate its difficulty. Anyone who learns undefined terms well deserves plenty of praise. And anyone who needs plenty of time to learn such abstruse concepts deserves patience. A parent who uses math words helps.

One of the advantages of being a college professor in a first grade classroom is that I don't lose face by saying to a child, "That is a very hard question you are asking—so hard, I don't really know the answer. I hope you will continue to think about it yourself. Right now, I hope you will concentrate on the things I *can* teach you."

How often are children accused of being stupid or lazy because they ask a profound mathematical question? If they believe their accuser, society loses big. They have learned the worst lesson a child can learn.

Materials and Activities

The challenge of teaching math is to seize its simplicity without denying its difficulty. Back in the thrifty days of my childhood, my parents taught me math with a pencil, scrap paper already used on one side that Dad brought home from work, and my brother's blocks. (My parents were liberated enough to encourage me to play with his toys, but nobody in those days considered giving a girl blocks of her own. If you want your daughters to be prepared to compete in the career marketplace of the twenty-first century, you should give them the same space-exploring and construction toys as a son.) Occasionally an uncle would give me a math book.

Although blocks, pencil, and paper are the essentials, you may be able to afford more. If you have the money, buying your child a computer and up-to-date software may provide an absorbing way to learn mathematics. Computers lack the social interaction of traditional games, however, and may be harder on the eyes and back. Computers and appropriate software are expensive and beyond the scope of this book.

BLOCKS AND PUZZLES

Blocks that lock together such as centimeter cubes, snap cubes, or multilinks are great for building creatively while subliminally exploring spatial relationships. Pattern Blocks and their ilk are great for two-dimensional fantasy and learning. Free play with these can teach much.

Puzzles of all types entice children and grown-ups into mathematics. Make up some. Scour your library for good puzzle books. If you can afford them, put puzzles temptingly around your home. Jigsaw puzzles can gradually become more complicated.

Some parents habitually go for long walks with their offspring, sometimes solving puzzles or math problems. One grandmother now remembers how her husband and son used to go for a walk after dinner and often came home elated that they had "solved it" together.

CALCULATORS

Most families now own an inexpensive calculator. Let your child play with it. The role of calculators in classrooms is still controversial, but it is clear that the recreational use of calculators opens new vistas.

Calculators keep grown-ups honest. Serious math can't be evaded after children get their hands on a calculator. If you say, "You can't take five from three," or "You can't divide seven by two," the calculator will contradict you. If a calculator can do these problems, why can't we? Today's child has access to numbers that were forbidden in pre-calculator times. Mathematical lying is not as easy as it once was.

"When you are older" is not an ideal response. The best time to answer a child's question is when it is asked, and the next best time is as soon as possible thereafter. Saying "When you are older," however, is better than claiming that people "can't" do the problem.

As your child learns to read, leave around books of riddles, puzzles, and games. *Family Math* has become a classic. It has a number of worthy successors. Biographies of mathematicians are appearing for grades 3–7. The catalogs whose 800 numbers are listed in Appendix C invite browsing. Buy math presents for your own children and others on your gift list.

Possible Responses to Embarrassing Questions

Suppose that your child subtracts $3 - 5$, and demands to know what -2 means. What might you say? It depends. On a farm, you might contemplate what happens if an ant that is 3 feet above ground crawls down 5 feet. In a city, you might consider where you would be if you start on East Third Street and drive 5 blocks west. Or you might draw a picture.

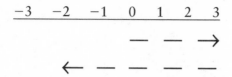

If your child divides seven by two, and asks about 3.5, your challenge is harder. You might take seven straight pretzels and suggest that your child split them fairly between two people. You might ask where the halfway point is to stop and rest if you were going to walk seven blocks. Or you might draw a picture:

These suggestions, however, give your child only a glimpse of the concept of three and a half. Why ".5?" Depending on what Meadow your child is in, you might talk about tenths, dividing something into ten parts, and showing that 5/10 is half of a whole. If your child knows about place value, you could talk about going backward from the hundreds place to the tens place to the ones place to the tenths place. Your child's comprehension may be fuzzy, but it doesn't matter too much. It isn't your job to judge your child, just to answer questions to the best of your ability.

If your child keeps prodding, you might (1) keep trying until you run out of time, (2) ask for a reprieve until tomorrow (and try to think of a better explanation overnight), (3) retreat to "When you are older," or (4) say, "I guess I don't completely understand. Shall we try to find someone who does?"

If your child pretends to understand and you suspect it's superficial, no problem. You've sown seeds that may someday sprout. There's no hurry, especially for ideas ahead of "schedule."

TIME LINES

Time lines are fun to sketch at home. They teach both math and history. One day my mother made a time line that showed the comparative times of Jesus, George Washington, and her own birth. She said that the dinosaurs were so far off the paper that we couldn't even draw them in.

I remember the strange swirl of emotions when she told me I had been alive such a short time that my birth was less than a pencil-point width from "Now." Everything I remembered had happened in such a short time! It seemed so long to me. My mother must have lived an impossibly long time. . . . Life must be endless. . . . (Then I realized not so many years later that time was indeed passing—and actually seems to accelerate. Life is not endless, but a precious adventure not to be squandered.)

MAPS

Map reading provides plenty of math problems. Drawings of a room or your home are the easiest to create, and next are local maps that include every street of your neighborhood. Your child can see the streets down which the family walks or rides in miniature. If it takes ten minutes to walk to school, which is one inch from your home on the map, how long would it take to walk to the dentist, whose office is three inches away? Such problems use multiplication and proportions and relate geometry to arithmetic.

State maps are more advanced. How far will we drive? How far to the next town? What is its name? I remember the thrill of being able to predict what the road signs would say just by looking at a piece of paper on my lap.

- Should we turn right or left at the next intersection?
- When will we get there?
- How can we use the map to predict when this ride will be over?

ALLOWANCE

Allowances provide a powerful incentive to learn arithmetic. They teach decimals. What is a reasonable weekly allowance? Why? These are good thought-provoking questions in practical math. If you and your child can decide together what the allowance should be, you can prevent those unpleasant begging scenes in stores. "If you want it, use your allowance. All gone? Buy it with next week's. Too expensive? Learn the power of saving."

One family gave an allowance to each of their children with rules about what fraction was to go to savings and what fraction to church. As the children grew older, their allowances grew and so did the amounts for savings and church. The children were expected to compute the correct amounts using their knowledge of fractions. The parents made sure they did so correctly.

PRAYER

Some mathematicians have told me that they often pray with students who come to their office in mathematical trouble, and it seems to make a big difference. If prayer is part of your family life, you may want to include math in your prayers. My own theology wouldn't encourage praying for right answers to specific problems, but praying for patience, hope, and wisdom seems acceptable.

One politician commented, "There will be prayer in the schools as long as there are math tests." It was meant as a joke, but the waiting for mathematical insight is indeed similar to waiting in prayer for those of us who do both. There are mathematicians in many religions who regard their mathematics as an expression of their religious belief.

Traditional Games

There are so many good new games that the best an inclusive book like this can do is mention some in passing and refer readers to the catalogs in Appendix C. Most cultures are full of children's mathematical games. If you want to steep your child in African math culture, for example, consult Claudia Zaslavsky's books. This section describes more of the games of my own childhood that, in retrospect, contributed to my mathematical life (see Chapter 6 for others).

RUMMY

Each player is dealt seven cards. The dealer then places most of the deck face down, but puts one card face up to begin the "discard pile." The goal of each player is to put down cards in sets of three or more. These three can be either (1) the same number or (2) a sequence in the same suit. Play rotates around the table.

Each player can either pick up the top card on the discard pile or the top card on the (face-down) deck. If a player has three cards in a set, they are put down face up. A player can add to these sets if another card in the sequence or the same number is picked. Whenever a player puts down a set or a card, that player can then draw again. Each hand continues to have seven cards. When a player has nothing else to play, the player puts a card from his or her hand face up on the discard pile.

The game ends when some player can't play any more. The person with the most cards showing face up wins.

This game teaches order of numbers and some strategy. However, it involves enough luck so people of different ages can enjoy playing it together.

For children learning arithmetic facts, you can make or buy special decks with the problems on them. If the answers are the same, the cards belong together. In other words, while practicing addition, 4 + 5 belongs with 6 + 3 and 1 + 8. You can play Concentration, Go Fish, or Rummy with such cards. They are also available commercially.

PARCHEESI

Parcheesi is an old-fashioned board game that requires more of the players than the simpler games for preschoolers. It teaches number recognition, addition facts through six (because each player tosses two dice and moves the number of spaces shown on both), and strategy. To play Parcheesi, you need a commercial game. If you don't own Parcheesi and are short on money, you're probably better off spending it on pattern blocks or base-ten blocks.

CHECKERS

Checkers is a game of pure skill between two players who take turns. It is played on the black squares of an 8-by-8, 64-square board. The red and black squares alternate. Each player has 12 checkers; one has 12 red checkers and the other has 12 black checkers. At the beginning the players sit at opposite sides of the board and put their 12 checkers on the black squares nearest them.

Players take turns playing. There are three kinds of plays:

1. A move forward. Each checker can move to an adjoining blank black space toward the other player. Each checker has at most two possibilities. At the beginning, 8 of the 12 checkers are completely blocked, and most remain blocked.

2. If an opponent's piece is in "front" of your piece in this sense, and there is a blank space beyond, you can jump over your opponent's piece and pick it up. If you land in a place where you can do it again, then you can make another jump—a "double jump." Sometimes people play that if you can jump, you must. By other rules, jumping is optional.

3. If a player gets a piece to the other side of the board safely, that checker becomes a "king." Another checker (one that has been taken by the opponent) is put on top of the new "king" to symbolize its power. A king can move in any direction to adjacent black squares.

The player who takes all the opponent's pieces first wins.

Even some kindergartners can play checkers, and it is still entertaining to many adults. When children become proficient, they are ready for chess, go, or owari, the complex traditional intellectual games of Europe, Asia, and Africa, respectively. All these games teach logic and concentration.

BATTLESHIP

Battleship is played on a 10-by-10 grid. These days, we can use graph paper; during my childhood, we drew the lines. Two players take turns guessing where each other's "ships" are. The first person to "sink" all the other's ships wins.

Before play begins, each player marks in a battleship (four adjacent squares), two cruisers (three adjacent squares), four destroyers (two adjacent squares), and four yachts (one square each). Adjacent squares can be diagonal to each other, and the multi-square ships can be in any configuration, as long as each square is adjacent to at least one other square in that ship. No ship may be adjacent to any other ship.

The grid is labeled like an x-y plane. Players take turns saying spaces on the grid. For example, when a player says "three, four," she or he is asking about the square three spaces to the right of the left border and four spaces up from the bottom. The response could be "no hit" if the opponent has nothing on that square, or "a hit on a. . . ." naming the type of ship. If there is a hit, the successful player continues to ask about other squares. When a player guesses wrong, it is the other player's turn.

A ship "sinks" when all of its squares have been guessed by the opponent. When a player succeeds in "sinking" all of the opponent's ships, that player wins the game.

Battleship teaches reasoning and the coordinates useful for algebra. However, it involves enough luck so that beginners can enjoy playing it with more experienced players. Many other games in the book *Family Math* help children explore and understand standard and nonstandard graphs, along with other vital math topics.

Solving Daily Problems

"How shall we schedule the day to get everything done?" That's a question every busy parent asks. It's a good math problem, using estimation, addition, and logic. How can we save time by doing errands near each other? By doing two things at once? Discussing these strategies with your child will teach basic arithmetic and problem solving—and just might garner some sympathy for your own challenges!

Any extended household project offers math problems. Talk about them with your child. Cooking "from scratch" obviously has many. Constructing, repairing, or rearranging anything involves math. If you are buying new shades or curtains, show your child how you measure the window. If you are buying any appliance, go out of your way to measure the place where you will put it and to discuss whether it will fit. (This might be a good precaution for other reasons, too!) Can you rearrange things better? Use scale drawings to help you plan; they will also teach your child geometric concepts.

Figuring out a store's change before the cash register does can be fun for children in mid-elementary school. Neighborhood mom-and-pop storekeepers used to teach the "missing addend" concept of subtraction while making change. They would start by saying the price of the purchase and then, while putting money into the customer's hand, state partial sums until they reached the total amount the customer had given. Now the cash register computes the change and maybe even spits it out. The only game left is seeing if we can estimate the total amount of change fast enough to catch the machine in a mistake.[5]

We can still discuss social problems with our children that teach addition, multiplication, geometry, and averaging:

- ▶ How many people live on this block?
- ▶ What shapes would we see if we could fly over this block?
- ▶ How much money do we spend on food in a typical week?
- ▶ How much is that per person?
- ▶ How much do we spend on food in a year?
- ▶ How does that compare with the average income in Brazil? In China?

Centuries-old math puzzles and brain teasers are still fun. In second grade my mother told me the following one:

Think of a number from one to ten.

Double it.

Add six.

Take half of your answer.

Subtract the original number.

Your answer is three!

My amazing mother! What a mind-reader! Now I know that this problem and similar problems have been used to "wow" children for a long, long time.

Why Does It Work?

When I asked my mother to explain how she knew my answer, she seized the opportunity to teach me a little algebra.

Call the number you thought of "x."

Doubling it, you get $2x$.

Adding six, you get $2x + 6$.

Taking half, you get $x + 3$.

Subtracting the original number, the answer is three.

The problem can also be understood using geometry:

———————	Think of a number.
——————— ———————	Double it.
——————— ———————- - - - - -	Add six.
——————— - - -	Take half.
- - -	Subtract the original number.

Algebra is not customary in U.S. elementary schools, but it is creeping in. Stevenson and Stigler's observers in Japan saw fourth graders happily chewing up such problems as, "If a class has 38 children and six more boys than girls, how many boys are there in the class?"[6] Just because you didn't have an opportunity to consider such problems until high school doesn't mean your child must be similarly deprived.

Work or Play? What Should a Family Try to Accomplish?

Parental support is essential, but what does that mean? If your child brings home math games from school as family homework, of course you will try to make time to play. You will try to use conversation, vocabulary, and problem-solving that encourages mathematics.

At home math should be fun. The teacher may dub math "work" at school, but at home it can be play. Math can masquerade as either work or play. Sometimes it's both at once.

Following directions is essential for school math. At home directions can be either work or play. While sometimes you *need* your child to follow directions, at other times you might ask playfully, "Can you draw a green circle above a red rectangle?" To make it a real game, you have to let your child make up instructions too. Can your child make up directions that you can follow? This role reversal can be great fun, and prepares children both for delving into real math problems and for that ultra-civilized custom of testing. Your goals for your child's schooling include both getting an education and beating the system. At least, mine did.

Your role in the primary grades should be governed by love and joy. It's much too soon for Tough Love. The primary years are the time to help your child recognize and revel in pleasures of the mind.

Who should know what? When? Our educational system is obsessed with judging children to make sure they are "normal." No wonder people are afraid to think! In their book, *Women in Power*,[7] Dorothy Cantor and Toni Bernay repeatedly emphasize that the top 25 women in U.S. politics, surveyed in about 1990, had one major thing in common: parents who loved them and kept telling them they could do anything they wanted. Try not to judge your child's achievement. That's someone else's job—if it's worth doing at all.

Yes, I believe that excellence is possible and that children, teachers, and the rest of us should reach for the stars. On the other hand, I clearly remember tidying up my knowledge of the multiplication tables in fifth grade. By today's standards I was two years behind, but who cares? In fourth grade I could figure out multiplication facts as I needed them and eventually, I became a math professor. That was good enough.

Nobody's consistent. My children knew their multiplication facts much earlier than expected and it's my fault. I really don't think I ever thought they *needed* to know them so early. However, asking them questions like "What is 6 times 5?" when we were driving around kept them from fighting with each other, and they seemed to enjoy it.

Many children can learn quickly when they are tiny, but early proficiency is not a predictor of later achievement. I know one man who didn't talk until he was three years old and later taught at Yale Drama School. My father knew a man who didn't talk until he was five and later earned a Ph.D. in chemistry and became an outstanding research chemist.

Both were white males born before 1940. The major advantage of early achievement today is that the schools won't "classify" your children as slow and use that as an excuse for denying them the advantages they need for high achievement later.

Individuality Is Okay: Elaine's oldest son had number sense when he was three years old, but the next, only thirteen months younger, didn't have it until he was eight. She believes she treated them both the same. The older one became a theoretical engineer and the second, who "couldn't add" for a long time, is now an accountant.

Fashions change. Most ten-year-olds today have the knowledge and skills that were taught at U.S. universities in the eighteenth century.[8] Many countries devised their curricula in the mid-twentieth century, when ours was already frozen into late nineteenth-century customs. (The NCTM *standards* were written to help our schools update their curricula. See Chapter 11.)

Please take the above disclaimers seriously. Children grow in many patterns. The culture keeps changing. It is virtually impossible to know what a person knows. Despite these caveats, there is a widespread and justified concern that children are not up to snuff in this country. How is a parent to know when to be really concerned?

It isn't easy. The problem is complicated by the fact that different people have different (and strong) opinions. Although there is less disagreement about the mathematics curriculum than there is in other fields, groups do dispute what should be learned and how to measure it.

Unpredicted Successes

Thomas Edison was refused an education in his public school on the grounds that he was retarded. His mother educated him at home. Albert Einstein received only C's in high school math. Colin Powell was an undistinguished student whose teenage friends now say they never would have dreamed their companion might become a national figure. People are not products; their future cannot be predicted.

What Math Should Your School Child Learn?

The next chapter (Chapter 10) presents topics that you may want to make sure your child learns because they tend to be neglected in U.S. schools. Chapters 11 and 12 discuss two types of achievement standards. Chapter 11 summarizes the recommendations of the 130,000-member National Council of Teachers of Mathematics. It focuses on skills and concepts that should be learned by fourth grade. Exactly when is unimportant.

Chapter 12 suggests facts and ideas to be learned in each grade. It reflects what is actually happening in good U.S. schools in the 1990s, but I hope parents and professionals don't let the floor become the ceiling. Also, the chapter advocates a reordering of some topics. The chapter concludes with a long section about patterns that underlie basic arithmetic but are rarely described geometrically.

Please don't view either Chapter 11 or 12 as a ceiling. If your child wants to fly further, fine! Eternal vigilance is the price of math proficiency in your child. These are vulnerable years. Collaboration between home and school is essential. Your child's mind is your concern, too, and should not be relegated to professionals. Children acquire their fundamental enthusiasms from beloved parents.

Chapter 10

Math Topics Your School May Not Teach But You Can

[The Japanese] view the weaknesses [in our educational system] as a serious threat to the economies of both the United States and Japan. . . . The Japanese delegation to the U.S.-Japan bilateral trade talks . . . argued that Japan cannot be held solely to blame for the enormous American trade deficit and that at least part of the problem must be traced to America's lack of an educated work force that is required for industrial excellence.

—Harold W. Stevenson and James W. Stigler[1]

Not having national curriculum standards doesn't render basic ideas any less basic. Mathematics can be viewed as a tower of ideas. An occasional small hole near the bottom is not ruinous, but if there are too many holes or if they are too big, the tower eventually topples.

Children who miss one of the basics at the bottom of the mathematical tower experience a crumbling of their competence and confidence later on. They can continue to build for a while with a missing cornerstone, but the hole affects everything that follows. This chapter helps parents spot and patch the most common holes in the mathematical tower.

Perhaps the missing topics have fallen through the cracks of curricula lacking unity. (Remember that old gag about a camel being a horse designed by a committee?) Anyway, older students claim they were never taught certain topics, and I have become convinced that these claims are not just due to their fallible memories, although that may be true for any particular sob story. The same holes keep appearing.

When fundamental mathematics is missing altogether from the curriculum of many U.S. schools, and families don't fill the gaps, their offspring are doomed to drive four-wheel cars with only three wheels. Today we have too many of those three-wheeled cars stumbling around.

When you learn may be as important as what you learn. If you don't learn multiplication until after you have struggled with place value, then you may acquire the habit of suppressing questions, which will stunt your later mathematical growth. Furthermore, you will never feel as comfortable with place value as you would have if you had understood three times ten before they thrust thirty upon you.

If problem solving and geometry are in the back of the book, these basic topics may get lost altogether. If children are taught only the algebraic strand of math without the geometric strand, their mathematical tower is unbalanced. They miss the connection to three-dimensional reality.

This chapter presents five often neglected basics: number sense, measuring (especially length), subtraction other than takeaway, the relationship of multiplication and area, and fractions.

Be alert for teachable moments. If you are yourself uncomfortable with these topics, you may want to seek more help to get "on top" of them. When you spot holes in your child's mathematical tower, use your own teaching techniques, mine, or others. Whatever works that is not abusive is fine. It is not fine to leave crippling gaps in your child's mathematical foundation.

Number Sense

Counting is only one way to understand numbers. Each number is an idea in itself. Children who must count to reach a number don't understand that number for itself. Children who play with dice or dominoes become friendly with small numbers, and don't have to count the number of dots on each face of a die.

Recognizing dots on dice is only one way to develop number sense. If you say, "I wonder how many people are in this room," when there are fewer than six people, your child has an opportunity to "help" you by scanning the room and seeing the number—with or without counting. "How many wheels are there on this truck?" (which invites doubling the number you can see) or "How many petals are on this flower?" are other questions that may focus your child's attention on small numbers without counting.

You will find many ways unrelated to counting to introduce your children to numbers. Many arts and crafts have implicit patterns. Often gardens are laid out in patterns. Exercises and games can teach numbers kinesthetically. Musical families can relate numbers to rhythm. Most nursery rhymes have a strong beat. For example, "Twinkle, Twinkle, Little Star" teaches four, and "Rock-a-bye Baby" teaches six. Keep your eyes and ears alert.

Although I much prefer direct observations in daily life to dots for *introducing* numbers, over the years I have created thousands of dot cards (usually while on the telephone) to wean primary grade children away from habitual counting. Too many stumble along with counting as a crutch. Before distributing the cards to teachers and children, I get mine laminated. Generally, I'm not very artistic, but I have fun with pretty colors and patterns, and the dot cards are attractive. Through them, children "see" math similarly to the way I do, either by just admiring the cards or by using them for Concentration or Rummy games. Parents and children also can make up dot cards with inexpensive self-stick dots on file cards. You too can enjoy the imaginative patterns that you and your child will make.

Rudy Rucker, in his fascinating book *Mind Tools: The Five Levels of Mathematical Reality*,[2] extends this dot way of thinking to larger numbers. For example, he describes 13 not just as 10 plus 3, but also as $3^2 + 2^2$ ($= 9 + 4$) and $3^2 + 3^1 + 3^0$ ($= 9 + 3 + 1$). Then he draws these numerical patterns geometrically using dots, respectively as:

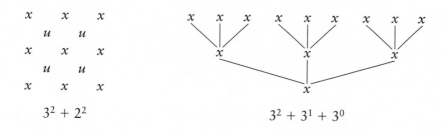

For a moment, ignore the numbers. Can you see the patterns in these pictures?

In the left picture, there is a large square of three dots in each of three rows. We call this "three squared," which is written 3^2 (and equals 3×3). Three squared equals nine. Among these dots, there is another square with two dots on a side or "two squared," which is written 2^2 (and equals 2×2). Two squared equals four. Adding nine and four, we get thirteen—and a special insight into 13.

In the right picture, there are three rows, one for each of the addends in the sum. The top row shows 3^2, or nine, in a different pattern from the pattern on the left. The second row shows $3^1 = 3$ and the bottom row $3^0 = 1$.

Can you and your child recognize quickly the number of dots in the following patterns? Talk about how many *different* ways you can use to see the answer.

      (a)   (b)   (c)   (d)

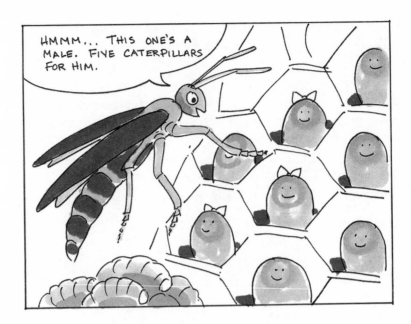

Apparently, some birds and insects have number sense, although they surely can't count. Typically, birds can tell the difference between the numbers two and three, but not between three and four. Some solitary wasps consistently put 24 caterpillars into each cell where they lay an egg so the wasp grub will have just enough food when it hatches. A *Genus Eumenus*, however, puts five caterpillars into each cell containing a male egg and ten into each cell with a female egg. (The females eventually grow much larger than their brothers.[3]) Some human cultures, in contrast, entered the twentieth century without the concept of four. Their language has only "one," "two," and "many." Four has to be taught.[4]

Math-loving people are accustomed to looking for patterns in the real world. We spend less time computing than other people, because we "see" numbers directly. Many young children display this math-loving behavior, but they need encouragement to continue.

It is sad when children are admonished to "stop guessing" because sometimes guessing is exactly what is needed. In general, guessing is fine, although not enough (see Chapter 4, Meadow 4, for another discussion of "numbers without counting").

When? No later than kindergarten.

Measuring Using Rulers

A few years ago, five Montclair State University students made a group appointment with a colleague of mine. When all five were present, they confessed to my colleague that none of them could use a ruler. Could he, please, show them how?

These were motivated students. They discussed their problem with each other, failed to find any student who could teach them, and sought out a professor. They weren't stupid. Montclair State is a good institution whose graduates become responsible professionals. These students had been terribly let down by the system.

One problem is that our country tried to switch to the metric system in 1974, amid considerable fanfare, but it didn't work. Metric measurements are far easier than our traditional U.S. system, but nothing is easy until you learn it. Our society now uses both systems, so neither is familiar enough to be comfortable. Alas, your poor child may not be taught either system in school.

If you know how to use a ruler, share that knowledge with your child. Most children love measuring things. Measure one side of a piece of paper and suggest that your child measure the other. What else would you and your child like to measure? (One family I know keeps a tape measure hidden in their living room to accommodate their second grader's frequent sudden urges to measure things.)

One thing both you and your child would find interesting is her or his height. If you're not too much of a neatnick, it's fun to mark each child's height at a particular place on the wall every three months or so. Label the heights. Everyone can see that growth is actually happening.

Guess measurements. Sometimes you and your child can guess together and then measure to see who was closer. Other times, you may want to just guess for the fun of guessing and talk about why you each made your guess. Maybe you don't have a ruler, or maybe you'd just rather not use it right now.

Length is an especially important concept because it underlies the number line, the mathematical idea that is basic to Meadows 7, 8, 9, and 10 (see Chapter 4). But other measurements are important too. You can find many opportunities to use numbers while measuring.

When? First grade or earlier.

"Subtraction Means Taking Away" Is Wrong!

My first class teaching elementary school as a college professor was a fifth grade. With an apprehensive tummy, I managed to get through my first half hour without serious mishap. Then one of the boys raised his hand.

"What is that word you keep using instead of takeaway?"

Perhaps nothing is more damaging in the current curriculum than statements such as "subtraction means taking away" (found in a currently popular book for parents[5]). Yes, takeaway is one interpretation of subtraction. To say, however, that "subtraction means taking away" is like saying:

- "Cooking means frying." Yes, frying is cooking, but what about baking, and steaming, and boiling?
- "Red fruit are strawberries." Yes, strawberries are a red fruit. But what about raspberries and apples?
- "Crawling is how children move themselves." Yes, crawling is one way that children move themselves. But what about walking and running?

- When you need three cups of flour for a recipe and your child has put in two, you subtract to find out how much more is needed. But you don't take away the flour you have already put in. You are searching for a *missing addend.*
- When you are comparing the brand-name price of $3 with the in-house price of $2, you may subtract. But you don't take away the in-house price. You are finding the *difference* between two quantities.
- If you have walked three blocks to grandma's and then two blocks home, you might subtract to see how many blocks you have to go. But none of that distance is going away. You have walked five blocks and will walk six by the time you reach home. You are using subtraction to measure *motion.*

All of these concepts are easy enough for any child who is capable of doing takeaway problems. Takeaway is an important concept. It certainly belongs in the curriculum. So do "missing addend," "difference," and "motion." Children find them all about equally difficult.

Children in the addition Meadow deserve all types of subtraction problems in their quest for new mathematical understanding. Harnessing children to takeaway deprives them of the vision needed for Leaping. Artificially separating the meanings of subtraction creates unnecessary barriers. Children need to learn four uses of subtraction simultaneously. Slip all four into your conversation:

Missing addend: If I have three cookies now, how many do I need to have five cookies?

0 0 0 ? ?

Takeaway: If I have five cookies, how many do I have if I give three away?

0 0 <u>0 0 0</u>

Difference: If Johnny has three cookies and Mary has five, what is the difference between the number of cookies they have? (We don't necessarily expect Johnny to get the same number as Mary or Mary to give some of hers to Johnny.)

0 0 0

0 0 0 0 0

Motion: If the joggers went five kilometers in one direction, turned around, and jogged three kilometers back along the same trail, how far are they from their starting point?

You can make up these problems as you go about your daily work:

▸ "Let's see. I want six boxes. I have four. Now, how many more do I need?" If a pregnant pause doesn't evoke an answer, you say brightly, "Oh, I need two more." Show the six boxes. ". . . because four plus two equal six." You can also say, ". . . because four and two are six." The boxes show that, for boxes at least, 4 + 2 = 6.

▸ "You have six dollars. If you spend four dollars on that ball, how much will be left?"

▸ "That line has six people. This one has only four. What is the difference between the number of people in this line and that one?"

▸ Or waiting for an elevator, you might say, "We're on the sixth floor. After we go down four floors, what floor will we be on?" Be sure it's a fun quiz. If your child wants you to give the answer, do so cheerfully. But don't be in a hurry. If your child is thinking, stand there patiently waiting for the elevator, presumably thinking about the problem yourself.

When? As soon as subtraction is introduced in school, but no later than second grade.

Area and Multiplication

Area, like length, is another form of measurement. Like length, area plays a more vital role in advanced mathematics than, say, temperature and weight. Length is one-dimensional; you can travel only along one line when you measure length.

Area measures two-dimensional things; we might say these things have both length and width. Or we might say they have length in two dimensions, one of which we choose to call "width." For beginners, it sounds confusing. For the rest of us, it's just the words that are confusing.

Areas themselves are not confusing in their simplest form. If _____ is one centimeter long, then

is 1 square centimeter. We can count how many squares there are in a figure. For example, the following rectangles each have an area of 6 square centimeters.

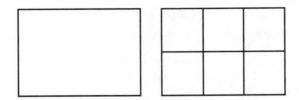

You can see that they are 3 centimeters wide and 2 centimeters high. In this case, the width and height are the two lengths. You multiply them (each is measured in centimeters) to get the rectangle's area (in square centimeters).

Let's do it again. The following rectangle is 12 square centimeters, because it is 4 centimeters wide and 3 centimeters high.

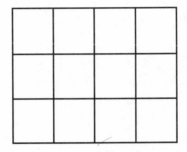

Eventually, you see that to get the area of any rectangle, you multiply. You can make rectangles with your children and have them find the areas. Of course, it becomes much harder if you leave the rectangles empty, so your child has to imagine the squares inside. Children enjoy using their imaginations. A "visual" child needs rectangles even more than a "left brained" child, but all children need to understand the relationship between rectangles and multiplication. Most find rectangles helpful and entertaining when memorizing times tables.

If you know how to find the area of any rectangle and any triangle, you can find the area of any two-dimensional figure with straight sides by cutting up the figure and adding its parts to find the total area. Thus, being able to find the area of a triangle is a basic geometry skill. I well remember my mother teaching it to me when I was in second grade. Take any triangle:

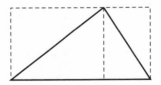

Wrap a rectangle around it. If you split the triangle down the middle, you can see that the rectangle has exactly twice the area of the triangle! In other words, any triangle has half the area of an enclosing rectangle. This always works. For obtuse triangles (those with an angle that is bigger than a right angle), you have to be clever in placing the rectangle around the triangle, but it still works. Drop the "altitude" from the vertex of the obtuse angle.

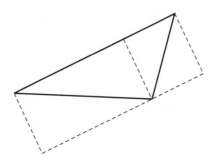

You and your child can have fun enclosing triangles in rectangles. Take turns making triangles for each other and drawing the enclosing rectangle. Then you and your child can have fun finding the areas of both rectangles and triangles. Try making up problems for each other using more complicated figures.

An understanding of the basic properties of multiplication follows from interpreting multiplication as rectangles.[6] For example, the Commutative Law is the property of multiplication that can be written $a \times b = b \times a$ **for any numbers** a **and** b. Often by itself, this law leaves people cold. It is easier to "see" geometrically. When you turn a rectangle on its side, the area is unchanged.

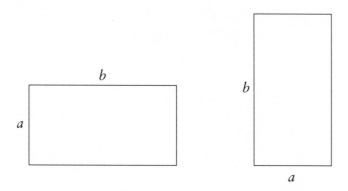

When? No later than third grade.[7]

Fractions

Fractions have many interpretations, and they are all important. If you don't understand fractions yourself, get someone to teach you before you try to teach your children. Written material isn't adequate. One reason is that fractions are complex; they involve many inter-related ideas. Another reason is that fractions are a basic concept; they cannot be defined in terms of simpler concepts. If you don't have a feel for fractions, much of this section may be unfathomable. It may be worth a try, or you may want to skip to the next chapter.

If music is part of your daily conversation, it may provide a vehicle for teaching your child fractions.

Definitions are doomed. One can read, "A fraction is a part of something," for first and second graders, and in the same series read, "A fraction is a part of one thing or a part of a group," for third graders.[8] A fraction *may* be a part of something or a part of a group, but it may also be a probability, a ratio, a proportion, or a division problem. I intensely dislike false and misleading definitions. Approach all elementary mathematical definitions warily. Because they use words, definitions must be based on more elementary concepts, but during the first decade of life, we must develop the most basic concepts nonverbally.

After children know small numbers, fractions arise naturally. The children don't have to be mature, as long as there is no expectation that they learn on a schedule. Surely, you talk about a half a piece of bread from time to time, or a half a sandwich.

Sandwiches and Fractions: When one mother made sandwiches for her daughter, she asked how many pieces the preschooler wanted. Dividing the sandwich into the chosen number of pieces, she would announce the fraction of the sandwich that she was handing to the child. "Here is one *third* of the sandwich. Here is the second *third* of the sandwich. Here is the third *third* of the sandwich." By the time her daughter began school, she knew the meaning of simple fractions.

Eventually, a young child wants to participate in the conversation and activity. As with all math, we must expect mistakes, acknowledge them occasionally, and communicate a strong message that mistakes are okay. It's also okay to simply ignore mistakes and change the subject. Keep it light.

Butter Sticks and Fractions: When Suzanne Granstrand bakes with her preschoolers, she tells them when she is filling a fraction of a measuring cup and points out the marking on the cup. Like all children, they appear interested well before she believes they understand what she is saying.

However, four-year-old Annemarie could answer some questions correctly. For example, when Suzanne gave her daughter a feeling of mathematical competence by asking, "Where shall I cut? We want a half stick of butter," Annemarie pointed to the appropriate spot on the butter stick.

After a few tries, a child understands enough about a half to notice that *that* fraction, at least, cannot be made by cutting anywhere on the stick, but only in one place. Sandwiches and butter sticks are only two of the many household items you can use to demonstrate fractions. Depending on your lifestyle, you might want to illustrate fractions using money, pizza, pieces of cake or pie, musical notes, or parts of the den that each child must clean up. Yes, fractions can describe a part of something. For example, the fraction ⅔ can represent what you get when you divide something into three parts and take two of them:

Two thirds can also be thought of as two things divided into three equal parts. An easy way to see that it's the same as the previous diagram is to divide each of the two things into three parts and select one third from each.

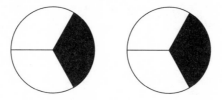

Laura and Mary were five- and six-year-old sisters in *Little House in the Big Woods.*[9] A neighbor they visited gave them each a cookie to take home. Laura and Mary each decided to save half of her cookie for their little sister at home. When they got home, Carrie had two halves. This seemed unfair to Laura and Mary, but they couldn't figure out what to do.

I remember posing this problem to my mother, who immediately said each girl should have eaten two thirds of her cookie. After contemplating this for a while, I decided my mother was brilliant. (See the picture just above to understand what happens if each girl eats two thirds of her cookie.) For years, I wondered how my mother could have thought this up so quickly. Now I wonder why I didn't ask her.

Think similarly about two meanings of ¾ and ⅝. Can you and your child draw pictures? As your child's understanding grows, you can draw diagrams of ⁷⁄₁₀, ¹¹⁄₁₆, and others. You could pose fraction problems for each other.

FRACTIONS AS NUMBERS ON THE NUMBER LINE

Just as you can visualize whole numbers as hanging on the number line (in Meadows 7 and 10 of Chapter 4), you can see fractions as living between the whole numbers on the number line. Just a few are shown below.

$$0 \quad \tfrac{1}{6} \quad \tfrac{1}{3} \quad \tfrac{1}{2} \quad \tfrac{2}{3} \quad \tfrac{5}{6} \quad 1 \quad \tfrac{5}{4} \quad \tfrac{3}{2} \quad \tfrac{7}{4} \quad 2$$

One odd thing about the fractions is that between any two fractions is another fraction! Your child and you may discuss that fact long and hard if one of you brings it up. There are indeed plenty of fractions!

FRACTIONS AS PROBABILITY

If you toss a die (singular of dice), you have a two thirds probability of getting a number that is three or larger. There are six equally likely possibilities: 1, 2, 3, 4, 5, and 6. Of these, four (3, 4, 5, and 6) are 3 or larger. Thus you have $\tfrac{4}{6} = \tfrac{2}{3}$ probability of getting three or larger. Fractions can be probabilities. This is a different meaning than "a part of something or a part of a group."

FRACTIONS AS RATIOS OR PROPORTIONS

Fractions can also be ratios. Ratios are one way of comparing one number to another. For example, if one number is ⅔ of another, we can say they are "in a ratio of two to three."

Ratios are often combined into what is called a proportion. Ratios and proportions can be visualized using rectangles. For example, the rectangle below has height 4 cm and width 6 cm. Thus the height is ⅔ the width; we can say "they are in a ratio of two to three." If we draw a diagonal in the rectangle, then any other rectangle that has one corner on its diagonal and two sides on sides of the original rectangle will also have its height two thirds of its width. The height and width of any two such rectangles can be, respectively, the numerator and denominator of two fractions. We can set them equal to each other (because they both equal the fraction ⅔), and we will have a proportion. Proportions are useful in many contexts; a "proportion" could be defined as a mathematical expression of the form $a/b = c/d$.

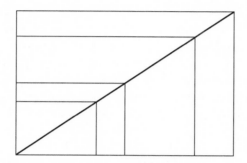

Knowing about rectangles is extremely useful for understanding proportions because any proportion can be visualized as two rectangles that are the same shape but different sizes. Looking at it another way, you can think of stretching and/or shrinking one rectangle that keeps its shape but changes its size.

FRACTIONS AS DIVISION

Just as ⅔ can be interpreted as two things divided into three parts, an arbitrary fraction "m/n" can be interpreted as "m" things divided into "n" parts. In other words, a fraction can be interpreted as a division problem. In algebra, division problems are usually written as fractions. Thus, it is helpful for children to think about the relationship between fractions and division early.

The ratio of nutrients in food or soil and of pollutants in many items is critical to understanding human health and environmental issues. Fractions as division and proportions are fundamental to these momentous matters of human survival. For example, if only organic matter can burn, if municipal waste is about one tenth inorganic matter (after glass and cans are removed), and if we burn half of the waste away, then the residue will be about one fifth inorganic matter. The figure below shows how this works. It is a diagram of a stick where one tenth of the stick is inorganic. If we burn away the half on the right, one fifth of what is left is inorganic. Thus the fraction of inorganic matter grows as we burn away organic matter. The particular numbers may vary, but the basic principle stands: burning anything concentrates the unburnables.

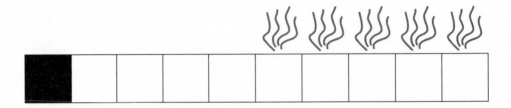

A very competent lawyer specializing in sludge told me that he couldn't understand the math involved in my question about heavy metals becoming more concentrated if water is removed from sewage. If a certain fraction of raw sewage is heavy metals, it seems obvious to me that if water (only) is removed, the resulting sludge will have a larger fraction of metals than the original sewage. The lawyer said he couldn't follow that math, although I used my best teaching skills to help him.

It is truly frightening that people who don't understand fractions as ratios and proportions are making far-reaching legal decisions about our environment. The same can be said about sharing economic resources, deciding on medical treatments, and reconstructing traffic accidents. Our country needs serious remedial work with adults. Fractions are basic to sharing and managing the world's resources on a finite planet. If adults don't understand fractions, the future of the human species looks bleak.[10]

When? First graders should learn the fractions ½, ⅓, and ¼. After that, it is crucial that children understand whatever they are doing with fractions. Too often fractions are taught as if they were nonsense.

What Your Child Should Know: Modern NCTM *Standards*

What has been most distressing since we released the Standards *documents is that our efforts to inform parents better have fallen short. . . . Some parents are concerned that because other, previously disenfranchised, students can now accomplish mathematics, this mathematics is not good enough for their children. This attitude has developed in spite of their mathematically promising students' having said in interviews that they have never been more challenged by, or more interested in, mathematics. We have to help those parents bridge their fears and encourage them to join hands in providing a solid mathematics education for all children.*

—*Jack Price, former NCTM president*[1]

What was good enough for Grandpa is not good enough for the twenty-first century. Elementary mathematics can be extremely profound, and the traditional U.S. curriculum has neglected these profundities. The *Curriculum and Evaluation Standards for School Mathematics,* published by the National Council of Teachers of Mathematics (NCTM) in 1989,[2] seeks to remedy this problem.

The *Standards* (as their friends refer to them) outline 13 areas of mathematics that children in grades K–4 should study. As this book is written, efforts to revise and update the 1989 *Standards* document are beginning. Although the NCTM recommendations make some people nervous, caring parents need to be aware of the modern professional attitudes toward what children should be learning.

There are many ways to cut a cake. Traditional standards often state specific facts or skills that a child should learn by a certain time. The NCTM philosophy is that specific facts and skills will come if a child acquires major understandings. My observation is that this is generally true, but some prodding may be needed for specific accomplishments.

A central focus of the NCTM is that rote facts and skills should not be forced on children in a way that prevents them from developing mathematical ability. Children need to explore novel, entertaining math ideas that will inspire them to memorize math facts because they realize it's "more efficient to know the facts than not."[3]

In the next chapter, we will "cut the cake" in a traditional manner. But here, we examine the thirteen NCTM standards for grades K–4.

The "Process" Standards

The NCTM *Curriculum and Evaluation Standards* have four "process" goals for students at all levels. Students should learn about mathematics as (1) Problem Solving, (2) Communication, (3) Reasoning, and (4) Connections. Solving problems is a major aspect of what mathematicians do, along with exploring patterns. Communication with others and within oneself (through both writing and speaking) is integral to mathematics. Reasoning is how we do most mathematics (although memory helps). Connections within mathematics and to applications in other fields keep math alive and "relevant." Through play and conversation, parents and other adults can help children with all of these "process" skills.

Often problem solving involves separating the math from the gift wrapping. If a problem says a woman bought a certain amount of tuna salad at the deli, don't worry about why it wasn't chicken salad or why she didn't buy it at the supermarket. While actually computing math answers, strip patterns of their settings. Concentrate on similarities; ignore differences.

To get the clues about what the similarities are, however, one needs to look for connections. Most riddles depend on surprising connections. I still remember my grandmother asking me a riddle from her own childhood: "What is black and white and red all over?" Her answer was "a newspaper," a play on "red" and "read."

A modern version is the following:

The masked men: You are at home. You make a right turn and three lefts and are confronted by two masked men. Who are they?[4]

Children love making up their own riddles. Nancy Litton's combined first and second grade in Piedmont, California, published their own mathematical riddle book. It includes this riddle: What has eight legs, two tails, four eyes, and one mane?[5]

Notice their mathematical creativity—far beyond mere counting and addition. Most children love riddles. Teaching children plenty of riddles gives them material to stump and entertain others and gives them confidence and joy in mathematics. You can find riddle books in almost any public library.

7-Up: Games require a longer attention span than riddles. One of my favorite oral games is 7-Up.[6] To play 7-Up, two players alternate saying either "one" or "two." When all the numbers said by both players add up to seven, the player who said the last number wins. People of all ages who don't already know the secret can enjoy 7-Up. It teaches problem solving, communication, reasoning, and connections.

> **In the Morning You Will Win:** One evening, while visiting the home of the five-year-old grandson of mathematician Emil Grosswald, I taught the boy the game of 7-Up. The adults listened for a while, but then Robert and I played unobserved. Robert seemed more puzzled than frustrated.
>
> "Why do you always win?" he asked me.
> "If you think about it in bed after the lights are out tonight, by morning you will know why."
> In the morning, the adults were having a jolly breakfast when Robert appeared.
> "You were right!" he announced to me.
> "Right about what?" The visit had had many highlights, and 7-Up was not on the top of my mind.
> "I know why you always won that game last evening. Let's play it again. I'll start." And, of course, he won. He used his reasoning and made the necessary connections to solve the problem after communicating well with me.

Computer scientist Michael Fellows observes that the "playground culture" of elementary school children "is rich with combinatorial games, with riddles and word-play, with informal discussion of infinity, space-time, and . . . jump rope, cat's cradle, and braids. These activities and puzzlements are in many ways closer to the spirit of mathematics than what is presented as 'mathematics' in the classroom."[7] The NCTM process standards, which are central to genuine mathematical growth, are sometimes taught better through the playground culture than traditional texts. Fortunately, this indigenous culture of mathematics is resurfacing in new school curricula and, along with brand new mathematics, is breathing life into old calculations that had almost lost their soul.

Number Sense, Estimation, Operation Concepts, and Computation

The next four standards are about arithmetic. They are called (5) Estimation, (6) Number Sense and Numeration, (7) Concepts of Whole Number Operations, and (8) Whole Number Computation. The last of these is what many folks associate with arithmetic, or even math.

Well-taught arithmetic has always helped children develop an understanding of individual numbers (number sense) and learn *how* to combine numbers using addition, subtraction, multiplication, and division (concepts of whole number operations). Drill and Kill, however, causes many children to emerge from elementary school without a good grasp of even arithmetic. A wholesome dose of traditional games, tastefully spiced with carefully chosen new games and computer games, should remedy that for *your* child.

Those of us raised in good traditional math programs (at home and at school) two generations ago learned enough number sense and operation concepts to make computation easy (but boring). Although we disliked the repetitive aspects of it (and carelessness and/or perceptual problems sometimes generated mistakes), we learned computation well enough so that we didn't need a chart telling us what "50% off" meant for purchases of $1.50 and $2.00. Such a chart hangs in a nearby store to help the young adults working there and prospective customers. Inadequate arithmetic skill has become a serious problem for many employers.

> **Number Sense and Warehouse Orders:** A warehouse foreman noticed that the workers putting labels on packages had trouble when they had to use numbered labels beginning with a number other than one. For example, if they were supposed to send ten packages and they started numbering with #2, they often ended the sequence with #12. They knew that if they started with #1, they should end with #10, but they didn't have the number sense to realize that beginning with #2 meant they should end with #11.
>
> Sending out eleven packages when only ten are paid for soon becomes expensive for the business. The foreman spent a great deal of his time checking that workers did not send an extra package to each customer.

Similar problems arise when understanding why eight-year-old children are in their ninth year (because one-year-olds are in their second year) or why 2001 A.D. is in the twenty-first (not twentieth) century—because the year 103 A.D. was in the second century. One way for children to develop number sense is to make up problems that interest them. This can be woven into daily family conversation. If it's not, children's lack of practice can be apparent. Once I asked a third grade class to make up math problems. Two responses were:

- ▶ If one class has ninety students and thirty pencils plus 50 dollars, how many in all?
- ▶ The car has 1,001 miles per hour. He drives 96 miles per minutes. How many did he drive in all?

Teaching rules instead of asking, "Why do you think that?" has been a continuing source of sad humor in the math community. One of the favorites decades ago was, "If it takes one woman nine months to make a baby, how long does it take three women to make a baby?" A modern version might be, "If I had one accident that cost $1,100, how much will five accidents cost?"

ESTIMATION

Estimation is a newly vital aspect of the arithmetic curriculum. Before your child uses a calculator to solve a problem, ask what the approximate answer will be. After the calculator has decreed its verdict, ask, "Does that make sense?" Calculators are here to stay, but they can make outrageous mistakes if betrayed by an errant finger or a dying battery. They need to be constantly checked for plausibility.

Yes, indeed, do encourage your child to play with a calculator! It's a great way to explore the concepts of arithmetic and to develop speed and accuracy. Already in the late 1970s the NCTM urged that every child have access to a calculator in every class. Too often, excluding calculators is an excuse for teaching computation instead of real mathematics. If you use math facts as pastimes during intellectually idle moments, *your* kids will learn mental math just fine. In some communities, parents have decided that since the basic facts are so much easier than most math, they should take responsibility for teaching math facts to their children and allow teachers to spend class time on harder math. It makes sense to me, as long as it doesn't cause tension at home.

Estimation has many uses that aren't new, just newly emphasized in the curriculum. As a parent, you can ask questions such as:

- About how much money should we take on today's shopping trip?
- About how much money did we spend?
- About how much longer will this trip last?
- About how many items are the people in front of us in this checkout line buying?
- About how many people are in this auditorium?
- About how many tulips are in that display?
- About how far did we go?
- If we have used a half tank of gas thus far, when should we refuel?
- Where should we park if we are going to shop at four stores in this mall?
- Does it matter if we intend to return to the car with a large package before the end of our shopping spree?

If you ask such questions, your child will pose them too. Asking them develops consumer and citizen awareness. Answering them develops number sense and skill at estimating. If you compute answers as you walk, ride, or wait in a line, you and your child will develop a broader understanding of whole number operations.

Geometry, Spatial Sense, and Measurement

The next two standards are (9) Geometry and Spatial Sense and (10) Measurement. You can teach these as you work in the kitchen, clean your home, rearrange your furniture, paint your rooms, make curtains, construct anything, plan your garden, report where you have been, or dress your child. Communicate about space. Make a model of your child's room and use it to try rearrangements of furniture. (A half inch on the paper for each foot in the room is a good proportion.) Share town and state maps with your child as you plan trips. Can you find a better route together?

Provide measuring cups and spoons in the sandbox.

Encourage your child to measure anything and everything. Then talk to each other about the discoveries. Two good openings for parents are "I wonder if . . ." and "Can you tell me about. . . ."

When teaching multiplication with base ten blocks, I start by having my students make a 14 from one "long" (which is ten units attached in a row) and four units. Then they put three of these side by side to see 3×14. It is easy to see (literally!) the answer from base ten blocks. Then we try 22×13, which most pairs of students can do in five or ten minutes with base ten blocks. (Individuals often become frustrated and give up.) It's fascinating to compare the resulting rectangle with the "long multiplication" of traditional elementary school. Areas of rectangles make multiplication much more understandable. For example, try showing that $(a + b)(c + d) = ac + ad + bc + bd$, when a, b, c, and d are any numbers, by using a sketch of rectangles.

If your child learns these concepts in the primary grades, later math will be much easier. NCTM *Standards*-based math programs include them. If your child seems clueless, perhaps you can help the school acquire base ten blocks and give the children a chance to explore rectangles.

The relationship between two and three dimensions fascinates primary grade children. When my husband's second grade teacher had the class cut up a piece of paper and make a box, he saw the potential and began making 3-D models of other things. His teacher encouraged him. His father then made a model of the home coal bin, and little Freddie was fascinated. He looks back at these events as precursors to his earning a doctorate in engineering.

Sometimes educational systems omit geometric concepts even more basic than rectangles, as the story about the five Montclair State students in the "Measuring Using Rulers" section of the previous chapter illustrates. Make sure *your* child knows about rulers and rectangles before leaving the primary grades. If you can't use a ruler or don't understand how areas of rectangles relate to multiplication (and this book isn't enough), ask your child's teacher or your Family Math leader, if you are involved in that program (see pp. 239–40), to explain. The teacher or leader will be flattered, and both you and your child will benefit mightily.

If nobody in your child's school responds to your questions, it's time to collect a group of friends and make an appointment with a nearby math professor. If one isn't handy, the math department chair of the high school that your neighbors' teenagers attend may be happy to have an opportunity to teach and share. Your inability to get a response in the local school will, very likely, be of great interest to him or her. Together, you can begin to move toward systemic change in your child's school and district.

Statistics and Probability, Fractions and Decimals, and Patterns and Relationships

The last three standards each have two parts: (11) Statistics and Probability, (12) Fractions and Decimals, and (13) Patterns and Relationships. The entire math curriculum of grades K–8 could justifiably be statistics and probability because everything else would be included. What portion of the class prefers pizza to hot dogs? Watches more than four hours of TV daily? Has more than one sibling? Has a pet in the home? These topics are interesting to youngsters and provide opportunities for counting, adding, graphing, fractions, and decimals.

Educator Susan Ohanian writes, "As I visited primary grade classrooms across the country, a glance at the student graphs on the walls showed me in which month most children were born, how many teeth they had lost, what color eyes they have, what type of shoe they were wearing on a given day, the types of pets they'd like to have, their favorite flavor of ice cream, and what vegetables they like. . . . Children think data analysis is great. . . . A six-year-old officially labeled as learning disabled was so intrigued by graphs that he started collecting lots of data. On his own, Peter kept track of what beverage his classmates chose for snack each day: milk, chocolate milk, or orange juice. He then devised his own system of graphing and reporting the results to the class."[8]

If your children are in an enlightened school, they too are becoming excited about data collection, analysis, and presentation. If not, do it at home. Statistics are indispensable for understanding our complicated world.

If your child has played games with dice, he or she has a head start when the class studies probability. However, theoretical probability begins with tossing coins, or *two*-sided items. Try tossing a coin or two or three with your child. Can you predict how many heads will occur if you toss one, two, or three coins 16 times? If you do it again, will you get the same result? If you engage in this activity repeatedly, what do you think you will discover?

Probability and statistics are a good way of learning about fractions. As mentioned at the conclusion of the previous chapter, the number of U.S. adults who never really learned fractions is depressing, because dividing things equitably becomes ever more important in a crowded world where almost everyone else on Earth is just the dial of a telephone away.

Whether fractions are interpreted as equal parts (of an item or group of items), or as division problems, or as probabilities, or as ratios, or as proportions, they are numbers. A fraction "lives" somewhere on the number line.

Where? I asked a fifth grade class in a high socioeconomic district where to place $\frac{1}{3}$ on the number line. One child thought it was a bit above 3, another around 5, and the third right on 3.

After considerable discussion the children finally concluded it was between 0 and 1. Even after that, it took a long while to realize it was nearer zero than one.

0	$\frac{1}{3}$	$\frac{2}{3}$	1	$\frac{4}{3}$	$\frac{5}{3}$	2

Then another child suggested we find where $\frac{1}{5}$ fits. Again, some suggested placement around 5 and around 3, but soon we were into the 0–1 interval. Where it should be in relation to $\frac{1}{3}$ took a long discussion. I folded paper into thirds and fifths and asked the children to compare the sizes and remember how relative size corresponds to placement on the line.

After we placed $\frac{1}{5}$ and $\frac{2}{5}$, a smart-aleck suggested $\frac{6}{15}$. Some children thought it was too big to fit on the line. After much discussion and folding of paper, they became convinced it belonged on the same spot as $\frac{2}{5}$. Apparently reducing fractions had never been explained to them as finding a new name for the same number.

0	$\frac{1}{5}$	$\frac{2}{5}$		1		2
		$\frac{6}{15}$				

The language of "reducing" fractions is especially unfortunate. The size of the fraction is not affected by "reducing it"—only the size of the numbers used to represent it. The numbers $\frac{2}{5}$ and $\frac{6}{15}$ are exactly the same size; they represent the same point on the number line.

To "reduce" $6/15$ is only to refer to it by a more familiar, friendly name. How different from the struggle of "reducing" a person's weight! How much does the emotional content of this inappropriate word interfere with children's learning to "reduce" fractions?

Worse: In a fifth grade class where I had already taught over ten satisfying lessons, I could not teach fractions at all. The children liked me, and were polite, but they knew I was daft to think there was a difference between the numerical expressions ½ and ⅓. Both meant "less than one," and that was that. Several weeks of trying were in vain. Anti-mathematics was already too deeply implanted.

If you don't make sure your children know the meanings of ½ and ⅓ by the end of third grade, they may never learn. With luck, they will marry someone like the wife of my superb graduate student mentioned on page 26, but parents are preferable to spouses when it comes to teaching fractions.

Better: Two years later I was in the same fifth grade teacher's class with the same lesson. This time the children quickly answered my questions and posed harder ones. The teacher and I stared at each other wide-eyed.

The difference? About a third of them had learned fractions with me two years earlier in third grade, and another group had learned them with a protegé of mine in fourth grade. About a half had learned what fractions really were from adults and had apparently taught the others. Fractions had become part of their culture. All the children knew they could solve and pose problems using fractions.

Decimals are another way of expressing fractions. Calculators use decimals. So do metric measurements. France uses the metric system, so decimals are used there throughout elementary school to study arithmetic. Fractions expressed with numerators and denominators are then introduced in the middle grades when the children study algebra. (Algebra requires numerators and denominators.)

Many kinds of patterns and relationships are appearing in the elementary school curriculum. Exploring them is both more fun and more effective than drilling on the old-fashioned topics, and children who perceive patterns nimbly learn traditional mathematics more easily. Patterns pervade topics in arithmetic, geometry, and logic.

> **Sequences:** Katrina's fourth grade teacher asked her pupils to continue the sequence 1, 2, 3, 4, 5, 4. . . .
>
> When her father posed this to me, I shuddered and said, "Oh, I hope it is 3, 2, 1!" As a math professor, I sometimes feel that my public image is at risk when people ask me about math problems from the primary grades. This time I could see several right answers, and I hoped I could guess the answer the local authority figure considered "right."
>
> Katrina's father sighed, "Well it is, but Katrina said 5, 6, 7, 8, 7. . . ." I grinned and nodded. That had occurred to me too.
>
> "Her teacher said she was wrong. But!" Her father's faith in humanity was renewed, "her classmates said she was right."
>
> He looked thoughtful. "The classmates she named were all boys. Finally, the teacher said, 'I guess the class is right.' Get that?" said the father. "Not that Katrina was right. The *class* was right!"

How do others' perceptions affect us? Katrina's father was worried about the math, about rightful credit, and about sexism. Patterns occur at many levels, and our perceptions affect what pattern is worth mentioning. As mathematics educator John Mason has put it, "Can you say what you're seeing?"[9]

Katrina is fortunate because her parents really care, and they believe she is lovable and capable. If she had been an equally bright child whose parents didn't care about math lessons, or who forgot to tell her parents about this incident, her self-esteem might have been tarnished. As it was, both she and her father were merely annoyed.

What Your Child Should Know: Traditional Checklist

I believe that if you have meaningful standards, . . . then you shouldn't be afraid to find out if they're learning the material, and you shouldn't be deterred by people saying this is cruel, this is unfair, or whatever they say . . . in the end, what the teachers and the principals and, more importantly even, what the parents and the children do is what really counts.

—President Bill Clinton[1]

—proposals for a new national system of achievement tests, at least as embodied in current proposals, seem to us quite unlikely to produce results that are either highly valid from a measurement point of view or promising from the standpoint of improving teaching and learning.

—Walter M. Haney, George F. Madaus, and Robert Lyons
National Commission on Testing and Public Policy[2]

As this book is written, there is a mass movement to create state academic standards coordinated across the country. The above quote from President Clinton was from a pep rally of state governors, who apparently are willing to collaborate. Since other countries have national standards, it seems like a possible goal. I am skeptical, not about reaching a mathematical consensus, but about politics and the intrinsic difficulty of measuring what people know, as reflected in the second quote. In social studies and English, there are serious academic problems as well as political and measurement problems.

Regardless of whether we have a standardized curriculum, the top goal is to preserve each child's "rage to learn." Situations are sure to arise in the next few decades that nobody foresees now. We can teach specific mathematical ideas while developing children's math power. This *must* happen so our children can become capable and confident problem-solving adults.

One reason for establishing national standards is the mobility of our population. All children have a right to construct arithmetic their own way, but they thrive best if teachers are not too inconvenienced. Some uniform expectations for children at each grade would have obvious advantages for children who move and their teachers.

If we do devise uniform national and/or state standards, each parent will surely want to borrow a copy from time to time. Even if we don't, home-schooling parents will want to look at their own state's current curriculum standards or framework. These policy documents should be available from your district's mathematics coordinator or from your state's Department of Education or math coalition.[3]

Current United States practices in math education have been dubbed "splintered" by the recently completed Third International Mathematics and Science Study. The study concludes that our schools are less focused than those in most countries and that we offer more math topics per year but with less depth. "This reflects the 'incremental' assembly line philosophy in American society . . . that encourages breaking complex learning down into simpler learning tasks assuming that, if students master all the small pieces, they will be able to put them together on their own."[4] The NCTM *Standards* attempt to help teachers help students put together the large ideas, but many of the traditional specific skills still need to be learned. Usually children absorb them as they grapple with more entertaining "real" math.

The splintered nature of U.S. math education may free parents (and teachers) to follow their own yearnings, but it also means that keeping an eye on cumulative skills is a good idea, since maybe nobody else is. Furthermore, as the guidelines of the National Association for the Education of Young Children say, "To develop . . . a sense of competence, primary-age children need to acquire the knowledge and skills recognized by our culture as important, foremost among which are the abilities to read and write and to calculate numerically."[5] Cultivating your child's computational skills is an important way to foster happy feelings.

Offering a specific set of standards is risky. Nevertheless, this chapter provides a summary of some traditional concepts and skills that are generally expected in grades K–5.[6] So that the floor does not become a ceiling, schools and teachers must include innovative material (see comments on pp. 109–10). It is both possible and desirable for children in the early grades to become familiar with the concepts of proof, infinity, variable, logic, induction, recursion, and computational complexity *if* their elders know something about these subjects. Dr. Mike Fellows, a strong advocate of giving challenging mathematical content to children, compares math to literature. When asked why children who will not be scientists need to know subjects like coloring and knot theory, he responds, "Does your child need to read *Charlotte's Web* or *Huckleberry Finn*? Does your child need to know about dinosaurs or outer space?" [7] He adds, "There seems to be a tendency for mathematics to be judged by stingy criteria that are rightfully not applied to other sciences, to history, or to literature." [8]

This chapter summarizes the "stingy" skills (the "phonics," to use Fellows' literature analogy) of mathematics. Obviously, one chapter cannot include all the essentials, so this chapter only offers an outline of the major concepts essential in the elementary grades. It is written for parents, not children, and is more specific than the NCTM *Standards,* but written from a mathematician's perspective.

Kindergarten

Children learn to count, at least to ten. The counting of things is often inaccurate, but accuracy is not nearly so important as the pleasure in counting. Number sense as explained in Chapters 4 and 10 should be developed; children should begin to see numbers independent of counting.

Recognition of squares, rectangles, triangles, and circles is desirable. Children should explore similar patterns in different settings: with manipulatives (blocks or cubes that teach patterns), drawing, singing, chanting, big-body motion, and clap-snapping. The more happy-go-lucky the problem solving, the better. [9]

First Grade

Addition is the theme. This presupposes all the Meadows through Meadow 5, as described in Chapter 4. If your child is not comfortable with numbers or numerals, try some of the practices suggested there and in Chapter 7, or turn to a book on math games.

Basic addition facts with sums up to ten are memorized. Subtraction can be learned incidentally by looking at ideas backwards. It need not be emphasized, but if subtraction is included, it should not be restricted to "takeaway" (see pp. 115–17).

Elementary data collection and display should be a major classroom activity since it may be *the* major application of mathematics today. If it isn't, you can help by doing some at home.

Simple multiplication is valuable, at least 2×3, 2×4, 2×5, and 3×3. Play with doubles by holding up the same number of fingers from each hand and placing the tips of the matching fingers from each hand together. If you hold up three fingers on each hand, your child can see that two threes (the number held up on each hand) is the same as three twos (the matching fingers).

Draw small rectangles to show multiplication facts. For example, the following rectangle shows that two threes (looking at the squares horizontally) is the same as three twos (looking at them vertically).

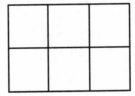

If your child's teacher introduces larger numbers, "skip-counting" by twos, fives, tens, and threes is another appropriate way to teach multiplication.

Make sure your child plays enough games to recognize the number of dots on dominoes and dice. The following patterns below are also useful:

Dr. James Callahan, a math leader in Vermont, believes that children in first grade should study math concepts using numbers less than ten. He urges that the complexities of writing numbers larger than nine should be postponed until second grade. He wants first graders to really explore subtraction, multiplication, division, geometry, and fractions with small numbers. This approach is viable, but most first grade curricula include larger numbers. If they do, your child should (for example) recognize how many dots there are in the following pattern:

First grade should be a time of exploring problems from many points of view. Alas, the research of Thomas Carpenter and James Moser at the University of Wisconsin shows that in many settings, children were better problem solvers at the beginning of first grade than the end.[10] This won't do! And it isn't necessary. If this is happening in your child's school, you need to be actively preserving your child's problem-solving ability—while considering what can be done to change the system.

Get even more puzzle, riddle, and game books. Your local library should have an abundant supply; if not, ask the library to buy some. Ask questions and pay close attention to any questions your child poses. Retaining your child's native creativity is vital in first grade. If you are home-schooling, use games and manipulatives from the fine catalogs listed in Appendix C.

If your child is learning math through manipulatives, problem solving, and extended conversation, count your blessings and praise your teacher and principal profusely! Don't take your good fortune for granted or it may disappear.

The Cognitively Guided Instruction (CGI) team of Elizabeth Fennema and Tom Carpenter in Madison, Wisconsin, has observed that first graders will concentrate on long, challenging problems if they haven't been taught that school is boring. Their team's first graders struggle with problems such as, "If I have 95 crackers, how many can each child in this class be given?" It may take first graders over 40 minutes to do such a problem, but they retain far more mathematics this way than by solving 20 two-minute problems. They also develop their ability to tolerate frustration.[11]

Try to make sure your child knows what a half and a third are no later than the end of first grade. Use the words "half" and "third" often at home, showing a half (and a third) of a glass of water, a piece of bread, and a cookie. Use rulers and tape measures in your child's presence. If your child is curious about what you are doing, measure lengths together.

Second Grade

All addition and subtraction facts are learned to a sum of 18. (In England, however, subtraction is memorized only to ten, and adjustments are made for higher numbers.) Children should understand subtraction as takeaway, missing addend, difference, and motion, as described on pages 115–17.

Traditionally, place value is taught, but most second grade teachers report their pupils don't "get" place value, no matter how hard they try. Either multiplication should be taught earlier, or place value should be postponed until multiplication is secure. How can you know what six tens and seven ones are if you don't understand the meaning of six tens or even six twos? Many U.S. children today actually learn place value in fourth grade, but that is unnecessarily late (and, therefore, hard on egos). If schools taught multiplication of small numbers using rectangles in first grade or early second grade, most children would be able to learn place value later in second grade.

Second graders enjoy contemplating negative numbers, multiplication, and division. These topics (at least multiplication) should be taught before "carrying and borrowing," or "regrouping," or "renaming," or whatever it is called this year. Then this (whatever-it-is-called) concept is easy to learn without much drill *if children understand the underlying ideas*. Traditionally, too much of second grade drills on this (name-changing) topic. Home-schoolers should follow an intellectually appropriate order so their children understand each topic before drilling on it.

Second graders also enjoy interpreting addition and subtraction using the number line. They should begin using rulers. They can learn where ½, ⅓, ⅔, ¼, and ¾ live on the number line.

0		¼ ⅓	½	⅔ ¾		1

Introduce symmetry and use it to explore simple fractions. Connect money (quarter and half dollar) and time (half hour and quarter hours) to fractional concepts. Name simple fractions using pieces of pizza, cake, or pie. Children learn all of these topics more easily by using rulers, measuring tapes, number lines, and a balance scale. Balance scales are available for about $5 and are more important to your child's mathematical development than a computer. Experimenting with the comparative weights of simple items like paper clips, dried ziti, and various types of dried beans and bottle caps will teach your child about multiplication, division, fractions, and proportions.[12]

Third Grade

Multiplication is traditionally introduced and memorization of multiplication facts begins. Children should be able to calculate them mentally up to 10×10 by the end of the year. Once this is done, it is time to explore addition and subtraction of two-, three-, and four-digit numbers. Numbers as large as you like can be studied all together if the children understand the ideas.

Third graders love large numbers. A reminder may be in order: A thousand thousands is a million. In the U.S. a thousand millions is a billion. In England, a million millions is a billion. See "'illions" on page 11 for a longer discussion of big numbers.

Fractions are often introduced. Make sure your child really understands halves, thirds, and fourths! Computation is premature until the *concepts* of fractions are thoroughly digested, but with the help of money, clock faces, pizzas, measuring, cooking, and geometry, now might be a good time to explore computation informally.

Mentioning division along with multiplication and fractions (sibling topics all) will set the stage for when your child must learn division seriously. If multiplication, division, and fractions are gradually begun in first grade, the harder concepts will be more accessible when they arrive.

The addition and subtraction "facts" should be memorized "cold," by the end of third grade. Multiplication "facts" are used regularly and should be easy to compute mentally, but no harm is done if it takes another couple of years before the child can recite them without missing a beat. Some teachers insist these facts be committed to memory in third grade, which is okay as long as the task is incidental to learning more serious and challenging math. Memorization should never be allowed to dominate math class.

Fourth Grade

Division should be learned, associated with areas, fractions, dividing into equal parts, proportions, ratio, and the old-fashioned "gazinta": How many times does 10 gazinta (go into) 40?

Mathematics in the primary grades should be far more than addition, subtraction, multiplication, and division. The book *Measuring Up*[13], published by the Mathematical Sciences Education Board, contains 13 prototype questions designed to test fourth graders on the mathematics recommended by the NCTM *Standards* for grades K–4. If you are home-schooling, you could ask your fourth grader to try these questions. They are fun to answer.

Fifth Grade

Fractions are explored in detail. If formulas run the show, your child may rightly rebel. Why learn nonsense? It is imperative that children have a feel for what fractions are before they delve deeply into adding, subtracting, multiplying, and dividing them. Even then, fractional computation is difficult. Extend sympathy.

Division of Fractions

One survey of teacher candidates[14] found that none of the fifteen elementary-education majors could make up a story problem to illustrate one and three quarters divided by one half. Furthermore, only four of the ten math majors preparing to teach high school math could do so. My husband and I decided to give you and your children a head start.

My problem: A mother has one and three quarter bananas. (She bought two, but a quarter had to be thrown away.) If she wants to give a half a banana to each child, how many children can she serve?

My husband's problem: Not quite remembering what a tire costs, a shopper took one and three quarters times as much money with him as the actual regular price when he went to buy a new one for my car. When he got there, he discovered the tires were on sale at half the regular price. Since his other tires were getting a bit bald, he decided to stock up. How many tires can he buy?

Bonus problem (looking at division another way): If it takes us an hour and three quarters to paint half of a room, how long will it take to paint the entire room?

Fifth grade is more inspiring when other mathematical topics provide "breathing room" amid the relentless challenge of fractions. Elementary algebra is common in fifth grade in other countries and increasingly popular in ours. Statistics, probability, geometry, number theory, functions, graph theory, game theory, and sophisticated approaches to measurement are sometimes offered. Our national curriculum is in flux, but most experts agree upgrading is needed. Many believe our students underachieve (compared to other countries) because they are not being challenged enough in math classes. They need harder math, not remedial math. The experience of many innovative teachers supports this belief.

Decimals were traditionally taught in sixth grade. How can they be postponed that late today? Calculators present them to any little child who can punch buttons. One divided by four is 0.25, according to my calculator. How does that relate to ¼? to ⁵/₂₀? to .25? to 0.25000000? to 25%? Six different ways of writing down the same concept!

Some people talk as if fractions and decimals and percentages were different from each other, but they are different only in the way that I am "Pat," and "Mother," and "Pat Kenschaft," and "Dr. Kenschaft," and "Ms. Kenschaft." I'm the same, no matter what you call me, just like one quarter (which is sometimes called "one fourth"). And just like one quarter, I do take on different roles. Does one quarter share my pleasure in variety?

Seize all opportunities to explain decimals to your children, for example, when discussing money and purchasing gasoline. Children in other nations have a much easier time learning about decimals because they use the metric system more often. Children here can,

but don't have to, learn about decimals when using rulers, pouring liquids, and planning distances. Decimals and percentages can and should be learned long before sixth grade.

Learning about fractions is difficult, whether you call them "*m/n*" or by decimal names or by percentage names. Fractions are worthy of significant school time. They are much harder than negative numbers, and should follow negative numbers in the curriculum. However, spending months learning alternative *names* (decimals and percentages) seems very odd to me. It suggests that someone making decisions isn't clear about the *concepts*.

Interpreting Arithmetic Laws Through Geometry

The commutative, associative, and distributive laws once were offered in college (until Sputnik jogged U.S. math education in 1957), then became middle school subjects, and are becoming fair game in elementary school. Smart teachers taught them to their youngsters for generations, even if they didn't call them by name. Today's children can enjoy naming these abstract concepts *if* they are taught geometrically.

The information in this section is not yet standard in elementary school, but children learn vocabulary more easily than older folk, and the topics illuminate primary grade arithmetic. Understanding these "laws" means less memorizing is needed.

Illustrating arithmetic with geometry makes understanding both easier. If you can read this section without too much anxiety, you may be able to make up geometric games to play with your child that will enhance arithmetic understanding. If not, don't worry. These topics are not *formally* part of the elementary school curriculum. They just lurk in the shadows. Being familiar with the ideas makes the primary grades easier.

COMMUTATIVE "LAWS"

Children notice the following sums are equal, perhaps using blocks or rods.

$$3 + 2 = 2 + 3 \qquad 6 + 4 = 4 + 6 \qquad 3 + 7 = 7 + 3 \qquad 2 + 5 = 5 + 2$$

They may notice that if we call any two numbers "*y*" and "*z*," then,

$$y + z = z + y$$

This is called the "Commutative Law of Addition." The word "commutative" comes from the same root as "commute," going back and forth.

As noted on page 138, young children can see that two threes are six by looking horizontally at the rows below. Looking vertically at the columns, they see that three twos are also six.

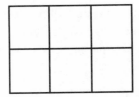

Similarly, from the following picture, they can see that three fours is the same as four threes.

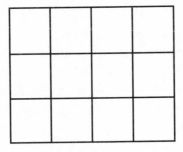

Eventually, a child sees that for any two numbers, called y and z,

$$y \times z = z \times y$$

This is called the "Commutative Law of Multiplication."

ASSOCIATIVE "LAWS"

Similarly, children may also notice the following pattern, while using blocks or rods to add three numbers at a time.

Adding $5 + 3 + 7$ can be pictured $\text{-----}\rightarrow\text{--}\rightarrow\text{-------}\rightarrow$ which can be added either

$$\text{-----}\rightarrow\text{--}\rightarrow \quad \text{-------}\rightarrow \qquad (5 + 3) + 7 = 8 + 7 = 15 \; or$$

$$\text{-----}\rightarrow \quad \text{--}\rightarrow\text{-------}\rightarrow \qquad 5 + (3 + 7) = 5 + 10 = 15$$

In general, we can write this fact $(w + y) + z = w + (y + z)$ and we call it the "Associative Law of Addition."

To teach the Associative Law of Multiplication, you need three dimensions. Old fashioned blocks will do, preferably in the shape of cubes. Better yet are multilinks, snap cubes, or centimeter cubes. Have your child make solid three-dimensional "boxes" (or "rectangular prisms") such as the one below and figure out how many cubes are in the solid.

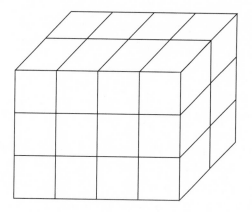

After children have played with these, they often recognize that they can find the number of cubes by multiplying each of the three sides, and it doesn't matter how you group the multiplication. For the above figure, we have $(3 \times 4) \times 2 = 3 \times (4 \times 2)$. In general

$$(w \times y) \times z = w \times (y \times z)$$

is the Associative Law of Multiplication.

DISTRIBUTIVE "LAW"

The "Distributive Law" connects addition and multiplication. Suppose, for example, we want to compute $(3 \times 2) + (3 \times 5)$. We can think of this as

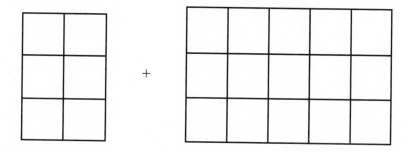

Since there are the same number (three) of twos as fives, we can rearrange these two rectangles into a single rectangle:

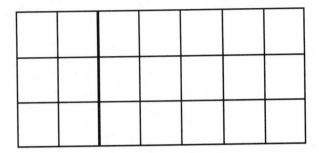

This gives us an easier way (or at least another way) of doing the original problem: $(3 \times 2) + (3 \times 5) = 6 + 15 = 21$ can be computed $3 \times (2 + 5) = 3 \times 7 = 21$.

You may also want to review the explanation of the Distributive Law on page 130 under "Geometry, Spatial Sense, and Measurements." There, rectangles show that $(a + b)(c + d) = ac + ad + bc + bd$ for any four numbers called a, b, c, and d.

Chapter

Getting Along with Your Child's Teachers

13

Wanted

College graduate with academic major (Master's Degree preferred). Excellent communication and leadership skills required. Challenging opportunity to serve 150 clients daily, developing up to five different products each day to meet their needs. This diversified job also allows employee to exercise typing, clerical, law enforcement and social work skills between assignments and after hours. Adaptability helpful, since suppliers cannot always deliver goods and support services on time. Typical work week 47 hours. Special nature of work precludes fringe benefits such as lunch and coffee breaks, but work has many intrinsic rewards. Starting salary $12,769 with a guarantee of $24,000 after only 14 years.

—Linda Darling-Hammond, 1984[1]

American teachers face formidable challenges.

Working Conditions

The working conditions of American teachers may be among the worst in the world. Although the Japanese school year includes more—and longer—days than ours, Japanese teachers spend fewer hours per year with children than U.S. teachers. Japanese teachers spend nine hours a day in school, but take nothing home at day's end. By national law, no teacher can teach more than 23 hours a week.[2] Each teacher has a desk in a teachers' room used for planning and collaboration.

In China a typical teacher teaches three hours per day and a new teacher may teach only one hour a day.[3] The Chinese school day has a three-hour break at midday when Chinese families, including the father, spend time together, eating, resting, and exercising.

American children are understandably restless. Harold Stevenson emphasizes that it's not the amount of time in class that matters, but how much time students *pay attention.* He claims we could dramatically and inexpensively improve U.S. math achievement by hiring playground supervisors to help children work off steam and develop social skills.[4] This would also give classroom teachers time to prepare lessons, grade tests, and do the inevitable paperwork that accompanies mass education without forcing the teachers to give boring activities to their pupils.

Public attitudes are also serious obstacles for U.S. teachers. When Japanese citizens are asked what causes high mathematics achievement, they answer, "Hard work." When U.S. citizens are asked this question, they respond, "Good teaching and high ability." If our children fail, they can believe it's beyond their control. Elsewhere the prevailing ethos puts the responsibility squarely on the child's own shoulders.

Social pressures on teachers are very different here than elsewhere. When Beijing parents were asked what was the most important characteristic of a good teacher, they answered, "Clarity. A good teacher makes clear explanations." To the same question Chicago parents responded, "Sensitivity. A good teacher is sensitive." Sensitivity promotes clarity, but placing nonacademic goals as top priority may befuddle both academic and nonacademic services in U.S. schools.

In most other countries the classroom becomes quiet when a teacher enters. In ours, people gossip about teachers' "ability to maintain discipline." In Japan, each day a different child in the class is responsible for the class's behavior. If the teacher tells the appointed child that the class is too noisy, that *child* must quiet the group down. Since the responsibility revolves, children shape up.[5] Teachers in this country find maintaining discipline both time consuming and exhausting. Acting as disciplinarians greatly detracts from being able to teach.

Teachers have virtually no access to money—even in wealthy districts. Teachers provide the actual education, but how much control do they have over where resources are allocated? Almost none. Furthermore, we have far fewer teachers per school personnel budget in the U.S. than in comparable countries. Most spend at least 60% (and some, 80%) of their school staff budgets on teachers. The average in the United States is only 43%.[6]

Too much money can be hazardous too. After receiving a $2 million grant for learning-disabled children, one of the country's most prestigious private schools diagnosed almost half of its children as learning disabled! Finally, the teachers rebelled, and the number plummeted from almost 200 to a half dozen.[7]

Special Needs Districts

In the less economically fortunate districts, nationwide problems are seriously compounded by lack of money. While most other countries have a national teachers' pay scale financed by national taxes with money distributed to schools on a per-student basis, our U.S. city and rural teachers often make less than two-thirds the salaries of nearby suburban districts—sometimes much less.[8] In the most inequitable states, the per-student expenditure in a wealthy community may be *five times* as much as that in a neighboring poor community.[9] Thorough studies contradict the assertion that money is not important.[10] Common sense does too. Why would the residents of wealthy districts pay high school taxes if they didn't believe it benefited their children? Why wouldn't it be even more important for children with fewer advantages at home?

The inequities go far beyond teacher compensation. Rugs, books, computers, overhead projectors, math manipulatives, scissors, and toilet paper, which are taken for granted in suburbs (and usually stay where they are stored), may be missing in nearby cities. When they arrive, they may be quickly stolen.

Samuel Freedman's book *Small Victories,*[11] documents the personal sacrifices and heroic efforts of teachers in one New York City public high school. Meanwhile, there are 6,000 people working in the New York City public schools' central administrative office although in the comparable office of New York's parochial schools, which serves about a fifth as many children, there are only 25 employees.[12] Having so many bureaucrats monitoring public school teachers indicates a distrust of U.S. public school teachers that seems unjustified to me. This distrust is pervasive, and is expensive for taxpayers.

> An urban teacher was taking a graduate course from a professor who was very impressed with her work.
>
> "Where do you teach?" the professor asked. She told him.
>
> "How long have you been there?"
>
> "Fifteen years."
>
> He looked aghast. "You could have gotten out by now!" he admonished her, oblivious to her sense of vocation and the great need for people who choose her path.

Respect

Students' lack of respect for teachers is one aspect of difficult working conditions, but it goes beyond that. My survey of math graduates of Montclair State who left teaching for other math-related careers revealed that they greatly increased their salaries by moving, but money was not the major reason for the switch. It was lack of respect from *adults*—

administrators and parents. Only one respondent was angered by student disrespect; the others reported that the youngsters were fine compared to the adults. U.S. teachers need and deserve more respect.[13]

Lack of structured collegiality is an enormous obstacle for U.S. teachers and, perhaps, another sign of disrespect for the profession. U.S. teachers are tucked away with their charges, unable to learn from each other and gather collective emotional strength. In Japan, teachers spend a major part of each day in the teachers' room with their private desk and other teachers. In the United States, teachers snatch brief moments with each other at best. When did your favorite teacher last sit in on a colleague's class?

Even less common is sharing teaching expertise with those in nearby schools or (heaven forbid!) in other districts. Try inviting a marvelous teacher in a nearby district to speak at your own school. See what obstacles are thrown up! Can you name any other profession where outstanding members are not encouraged to shine their light widely? Teachers are practically forbidden from using their most valuable resource—each other.

Class Interruptions

Lack of control over class time is another complaint of U.S. teachers. During 80 hours of observation in Chinese fifth grades, there were *no* interruptions; 47% of comparable U.S. lessons were interrupted.[14] The best-planned lesson may be continually interrupted by the intercom's insistent announcements, chastisements, and requests. This abomination seems to be everywhere, whatever the socioeconomic status. It is especially destructive in mathematics, where concentration is essential.

"Pull-outs" are another plague. Perceived disabilities have been routinely "remedied" by taking children, even very young, from their homey classroom to a special program. A teacher has her entire class for only short periods. (One third grade teacher told me that six different special teachers were "pulling out" some of her class at different times.) No matter how exciting, a lesson cannot "run over." Long, integrated lessons are impossible.

Only impeccable, time-consuming planning might prevent children from missing crucial work. Most children end up with gaps, which are especially pernicious in cumulative subjects such as foreign languages and math. Those with perceived handicaps are the most likely to lose crucial continuity. Enrichment programs also tend to "pull out" children from their classroom teacher, but people assume that these children are more capable of making up the missed lessons.

Furthermore, the reward system devalues classroom teaching by promoting teachers rated the best to be pull-out teachers. It takes courage to stay in the classroom when offered a pull-out job. The pay may not change, but the status is higher, the classes smaller, and, as one pull-out teacher told me, "You have time to go to the bathroom."

Math-Specific Inadequacies

All of the above affect every subject. Furthermore, U.S. elementary school and middle school teachers rarely get a decent mathematical background. Indeed, they may not be taught what mathematics is.

Twenty-four states have no requirements at all in mathematics for prospective elementary school teachers. Most of the others have only a minimal course or a very easy exam.[15] Nevertheless, "Roughly one quarter of newly hired American Teachers lack the qualifications for their jobs. More than 12% of new hires enter the classroom without any formal training at all, and another 14% arrive without fully meeting state standards. Although no state will permit a person to write wills, practice medicine, fix plumbing, or style hair without completing training and passing an examination, more than forty states allow districts to hire teachers who have not met these basic requirements."[16]

Even by third grade, teachers tell me some of their pupils know more math than they do. It takes a noble adult to avoid being resentful! Parents can help by being sympathetic, not indignant. It's not an individual teacher's fault.

One third grade teacher responded to my honest compliments by saying, "I really know third grade mathematics now. It's taken lots of work, but I really believe I know third grade mathematics. If they ever give me fourth grade, it will take a lot more work, but now I know I *could* learn fourth grade mathematics."

Some U.S. teachers are even less prepared. In her doctoral research, Diane Azin discovered that four out of fifty prospective elementary school teachers did not understand the concept of "two times three" *after* taking their undergraduate mathematics courses.[17] Throughout its 1996 report, the National Commission of Teaching and America's Future makes a strong plea for raising the standards of teachers' academic knowledge, especially in mathematics and science, but this will require far more public support than is now apparent.[18]

Practically no teachers want to harm their pupils, but it's impossible to teach something you don't know, and frightening to try. Teachers' frantic schedules prevent them from teaching themselves math and science in their "spare" time. Expecting them to do so is impractical and cruel. Alas, teachers in many school systems are not allowed, and certainly not encouraged, to increase their professional competence by attending workshops, conferences, and courses designed to help them become more inspired, knowledgeable, and effective. The paperwork, pleading, and procedures required from teachers who want to improve themselves is depressing.

Effective pattern exploration and problem solving require that teachers feel no shame when their students pose problems that they can't answer immediately. Indeed, such students can bring great pleasure to a confident teacher. However, teachers need to know the routine subject matter and must be able to solve most problems that students pose. They need:

1. Special summer and after-school programs to shore up their faulty math backgrounds, and extra pay for participating in such programs. It isn't their fault that states and universities don't require them (or even allow them, due to competing requirements) to learn adequate mathematics during their student years;

2. At least one math-loving teacher in the same building who welcomes questions;

3. Access to an occasionally visiting math specialist; and

4. Access to the "Information Superhighway," where computer communication with peers and leaders around the world can provide inspiration, ideas, and answers.

Even worse than inadequate teacher preparation are the attitudes and regulations of educational leaders (administrators and professors) who believe that mathematics is mere computation and actively oppose the learning and teaching of real mathematics. Indeed, some administrators go to considerable lengths to "protect" sensitive children from teachers who expect high mathematical standards.

Peak math experiences (Leaps across the Brooks of Chapter 3) can't be scheduled into a teacher's plan book. Administrators' policies sometimes damage teachers' professional freedom that is essential to "tend the garden" effectively. If a teacher is required to give an "anticipatory set" before the lesson, the children may be deprived of wonder. If the teacher articulates "closure" before the children have arrived at personal insight, they may be denied the excitement of discovery that leads to understanding and retention. In effect, a teacher's "closure" has revealed the end of the mystery before the child has read the book. It can take away the excitement of the discovery experience, and the children are not likely to look back on it with enough personal connection to retain it.

Well-prepared, well-supported teachers can stimulate their charges to gain several years of mathematical growth in one year. Administrators should allow them to do so, and not require rigid adherence to a teaching formula that stultifies.

How Do Teachers Cope?

I am continually awed by the kindness and generosity that teachers give to pupils, even in miserable working conditions. How do they cope?

First, the pleasure of math fades. In Chicago only a third of the teachers told Stevenson and Stigler's team that math was their favorite subject to teach.[19] In Beijing, China, where the elementary school teachers are also almost all women, 62% said they found math the most fun to teach.[20] If teachers don't enjoy teaching math, children are less likely to enjoy learning it.

A few teachers find refuge in power trips. Theirs are the classrooms of Drill and Kill cruelty.

Others have been brainwashed to believe Drill and Kill is good for children. The children's lack of progress is attributed to a problem with the children's ability, behavior, or culture. Such teachers don't realize they are stabbing the very heart of learning—the passion to learn. They mean no harm, but they teach children that "mathematics" is hateful, remote, and lifeless. A good summer program can revolutionize the mathematics teaching of such teachers.

Many other teachers simply cut math class short. The research of Stevenson and Stigler indicates that language and math are given about the same amount of time in Japanese and Chinese fifth grade classrooms, but U.S. fifth grade teachers spend over twice as much time teaching language as math.[21] A good summer program can revolutionize these teachers too.

No education is preferable to anti-education. Formal instruction is more expendable than health care or housing. Ironically, there are laws mandating education but none that mandate health care or housing. The result is a *pretense* of education where children are subjected to worse than nothing. Children running free can learn plenty, especially if their formal education is good when it happens.

I believe the cheapest way to improve U.S. children's math achievement would be to eliminate all "math" instruction until we have eager, well-prepared teachers. Children would teach themselves more math than they learn now from ill-prepared, math-anxious, overworked teachers. Research indicates that some teachers may be taking this approach on their own initiative.[22]

What Can Parents Do?

Whatever your child's teacher's problems, style, and expectations, you owe it to both your child and the teacher to be an active ally during this crucial period. Your child's attitudes, habits, and self-image are being established. They are too important to entrust solely to a teacher, who has all the above challenges, as well as other children, school politics, and relentless paperwork to deal with.

What can you do to help your child's teacher against such odds? Obviously, it depends somewhat on you, the child, the teacher, the school, and the community. However, there may be some guidelines:

1. Knowledge is power. Learn about your child's school by going to every possible PTA meeting and parent conference. If you can, visit your child's class occasionally, preferably both announced and unannounced.

2. Praise is our most underused resource. When your child's teacher does anything you like, praise! This is especially important when you are also dishing out criticism.

Learning in Many Gardens

You can tell when learning is going on in a classroom. It doesn't take special skill. You can see the excitement in the children and hear their involvement with ideas. You become involved yourself. You watch the unfolding of human thought. You don't need to give a test.

It's hard to describe the clues.

▶ Some learning environments are pin-drop silent, but once I felt I couldn't stand the noise of an enthralled class, despite its rugs. Most learning classes are between pin-drop silence and a roar, but both extremes are possible.

▶ Some learning classes are teacher-centered. Others are based on cooperative learning. Others have children working alone, sometimes on similar tasks, sometimes on very different activities.

▶ Sometimes the teacher is all over the place. Sometimes she stays put.

▶ Sometimes the teacher can predict what the day's plans are. Sometimes the children can steer activities in unplanned directions.

▶ Sometimes the subjects are highly integrated. Sometimes it's easy to tell when one lesson ends and another begins.

The more I visit many different classrooms, the less I believe any one educational philosophy is the "right" way. There are many ways to share facts, ideas, and inspiration. However, teachers and parents must believe in the approach being used.

The essentials are that the teacher is kind, knowledgeable, and free enough to implement his or her dream. You can tell when this is happening. I promise you. You can also tell when it isn't. It's dreadfully depressing. Of course, there are many classrooms that are neither inspiring nor depressing. Their teachers need help and freedom.

If your child's elementary school teacher is well prepared mathematically, she may be walking a political tightrope. Ordinary jealousies are compounded by the hazards of social change. Some people consider good math teaching seditious. Praise and support good teaching.

3. Regard your children's school activities as their work. Inquire about them. Show that you take them seriously. Provide a quiet time and place for homework.

4. Support the work ethos. Expect your child to work hard and often enjoy it. While your child is young, this is easy, but if a child is fed intellectual pabulum for a while, that child may resist gourmet food later.

5. Traditional parents always upheld teachers' authority; this is still desirable whenever possible. Children need their authority figures to collaborate. Try to avoid criticizing the teacher in your young child's presence. You can listen with the therapist's "um."

Research indicates that before 1960, "Parents might disagree with a teacher's style or perhaps look forward to the day when their child would have a different teacher, but their stance toward their child would usually be, 'OK, he [or she] is not the best teacher, but *he [or she] is the teacher and you have to listen.*' . . . the implicit consensus was that inside the classroom the teacher reigned."[23] Teachers, children, and parents might all be happier if this attitude were more common now.

You can support the teacher without having similar expectations of your own. All adults in authority need to maintain sufficient discipline to enjoy the children in their care, but the number of children and legal complexities make teachers' needs somewhat different from parents'.

6. Teach your child tact (or is it politics?). Help your child figure out how to please the teacher without compromising his or her own integrity.

Let's Not Tell the Teacher: While I was "helping" my parents do the dishes in second grade, my parents asked the perennial parental question of what had I done in school that day.

"Miss Mitchell told us that you can't subtract five from three," I reported.

"She *did*?!" My parents looked at each other with surprise and head-shaking. Of course, I wanted to know what was wrong. They explained the concept of negative numbers (see pages 45 and 101).

I was delighted. "Tomorrow, I will tell Miss Mitchell that we *can* subtract five from three," I announced with glee.

Again, my parents looked at each other in consternation. Again, there was much head-shaking. Helping other people learn was a positive value in our home, and I was puzzled. Shouldn't I help Miss Mitchell? My father looked appealingly to my mother.

"No, I don't think that's a good idea," she began slowly. "It might make Miss Mitchell feel bad." I'm sure I looked skeptical; I was *glad* when people taught me.

"Maybe she has a good reason for doing it her way." My mother brightened. "Maybe she thinks it would confuse the other children. That's probably it! She doesn't want to talk about negative numbers right now. I don't think it would be a good idea for you to bring it up."

Although I now know second grade teachers who find that their pupils easily learn and enjoy negative numbers, I think my mother's politics were right. It's threatening enough to have parents helping (i.e., correcting?) teachers. It's worse when children do it.

7. Can you help your child's teacher? Some teachers recruit help from parents. If you can respond, all the children in the class will benefit, and you and your child will have special psychic rewards.[24]

If you are tempted to *offer* help to your child's teacher, you must do so tactfully. Most teachers welcome such offers, but not all. Some teachers believe their classroom is their castle.

Most teachers really want to learn. Some parents with special backgrounds occasionally teach their offspring's class. Computer scientist Dr. Michael Fellows actually involved his daughter's second grade class in tackling heretofore unsolved problems. Some "open" math problems are accessible to young children, and they (like everyone else) are excited to work on problems for which nobody knows the answer—yet! Since then, Dr. Fellows has published articles and a book to help teachers and parents bring live mathematics to primary school classrooms.[25]

When There Are Problems

Sometimes, despite everyone's best intentions, a child really dislikes math class. What can you do?

1. Listen to your child. Give your child opportunities to vent feelings at home. If you feel inclined to give "obvious" advice, *discuss* it with your child. In general, do not give orders on negotiable matters. (You need to conserve your order-giving powers, which are skimpy enough in our culture, for nonnegotiable issues.)

2. Encourage your child to tell you what he or she likes about the teacher. What is going *right*?

3. Ask your child questions as gently as you can.

▸ Are there ways your child and the teacher can communicate more, better, or less?

▸ Does the teacher welcome offers of participation via hand-raising? Or is she or he intimidated by too much participation?

▸ Is the teacher impressed when a child makes corrections and extra contributions? Or threatened? Is eye-rolling with other smart children preferable to correcting the teacher?

▸ Does the teacher like praise from the children? Or does it seem to make the teacher feel judged?

▸ How does the teacher feel about jokes in class?

▸ Can your child find ways to help the teacher that might soothe the emotional wrinkles?

These questions should be explored very gently. Try not to implant fears or bitterness where it isn't justified, but acknowledge negative feelings so you and your child can handle them safely, preferably out of sight of the teacher.

4. Some of what a teacher wants may be merely silly, not destructive. Can you help your child comply faster, and thus less painfully? For example, can you figure out ways to sweeten an unpleasant homework assignment: Background music? TV? A conversation? Quiet company?

5. Listen to the teacher. Teaching is a difficult, undersupported, underrespected, vital vocation. Your child may not be an angel.

6. Discuss multiculturism with your child and the importance of adjusting to various styles. It may be a valuable lesson for your child to learn to get along with this teacher precisely *because* the teacher is so different from you.

If the problems are serious, the line between affirming one's child and undermining the teacher can be murky. To be able to give due respect without undue reverence is a valuable lifelong skill. All people have flaws, and teachers are people.

Children often have complaints about their teachers. If you listen to your child's complaints, you may realize that your child's behavior irritates the teacher in the same ways that your child sometimes bugs you. Your child (being a person) isn't perfect, and it is important to try to have sympathy with the teacher's responses, even if you don't tell your child about your sympathy.

7. Consider whether your child's teacher has legitimate discipline expectations that annoy your child. When your child complains about a teacher's attitude, try to help your child figure out what the teacher's view may be. Teachers are not always right, but they often are, and your child should make a serious effort to adapt to various types of people.

8. Suppose the teacher clearly doesn't know the math and is resorting to the methods described in Chapters 16 and 17?

Nobody's perfect. If your child loves the teacher, and you provide some real math at home, your child's mathematical ability probably won't be ruined. Mine wasn't. However, if your child hates "math" class, read carefully to the end of this book and plan a strategy, conferring with as many trusted friends and professionals as you can.

It helps if you don't alienate the teacher. If it seems prudent, go have a chat. You might bring this book along, and leave a copy. Read carefully the last third of this book yourself.

Tell your child that classroom peace may require capitulation to repetitive drill, but there's more to math than that. Getting along in one's own culture has merit—a topic you can discuss with children at a remarkably young age.

9. Continue alternative mathematics activities at home. If your child has already been taught to hate "math," call them puzzles or games. It is better to slip them in early enough so that your child learns that some math is fun and doesn't develop hatred for the word "math."

WHAT IF YOUR CHILD IS STILL MATHEMATICALLY MISERABLE?

If it's "only" math, keep trying. Reread this book; use the books in the bibliography; consult continually with professionals and your friends. Be alert, however, to the possibility that negative feelings about one subject may affect others.

If math is "the tip of the iceberg," more drastic action may be needed. If many or all of the parents are upset, make sure you are heard by the administration—first, the principal, then, if necessary, the superintendent, and finally, as a last resort, your district's Board of Education. Don't be intimidated. Listen carefully to the teacher and the administration, but insist on your right to understand. If the problem is that "new fangled" approaches are being used, parents should be part of the process. You *can* understand.

If the administration doesn't defend the teacher and many parents believe their children are being ill served, remember that incompetence is a legitimate reason to remove tenure. There are a few (very few) teachers who should be relieved of their responsibilities. The public is getting impatient with a court system that allows genuinely incompetent teachers to remain indefinitely in the classroom. We need strong parents to break the logjam.

Much more likely, your child and the teacher are a mismatch. Personally, I believe no teacher should have to put up with a child who really hates her or him. It is impractical to expect any teacher to be suitable for every child. A serious mismatch may be especially destructive in mathematics, where personal insight is basic.

If your child *really* hates a teacher, try to find another alternative. I'm not suggesting that every year is a popularity contest. Your child has to learn to get along with people who aren't perfectly charming. Both your child and your community deserve your best efforts to help your child get along with a less-than-perfect teacher. However, there *are* times for flight. I did it once (as reported in the next chapter) and I'm glad I did. The teacher was fine for many children, but not for my child.

Where public schools allow more freedom for principals and teachers and some choice for parents, children have options. Many school districts allow choice within the district. If our state laws mandating universal education are to fulfill the intentions of their originators, children and their parents should have teachers they respect. Surely there are appropriate teachers for your child within commuting distance. Does the law ensuring universal education in your state entitle your child to one? [26]

The Fifth Grade Crisis

14

[Grandma] understood many things that are barely recognized in the wider educational world even today. For example, she realized that arithmetic is injurious to young minds and so, after I had learned my tables, she taught me algebra. . . . She thought that . . . drill was stultifying. The result was that I was not well-drilled in geography or spelling. But I learned to observe the world around me. . . .

—Margaret Mead, *(reflecting on her own education at home)* [1]

How many American children go through serious mathematical crises around fifth grade? I know of no formal studies, but the anecdotal evidence is sobering. The minds of many ten-year-olds are being shamefully destroyed. Of course, the "fifth grade crisis" may happen in third, fourth, sixth, seventh, or eighth grade.

If you have been implementing the spirit of this book at home, you and your child are more likely to weather the crisis with only minor scars. Children subjected to torture in the name of "mathematics" can achieve mathematically if they realize *and accept* the nonsense of the situation. However, action may be needed.

A "Fifth Grade Crisis" occurs when your child experiences one or more of the following symptoms of trouble:

- ▶ more than ten homework "problems" in a typical homework assignment
- ▶ excruciating boredom during mathematics lessons
- ▶ social isolation mathematically
- ▶ intellectual stagnation

Mathematics homework problems should be sufficiently challenging that only a few can be done in one evening. If the "problems" can be done quickly, they are not problems; they are only exercises. If some are quick review, that is okay—but new challenges can't be hurried.

Some class time should be spent on group activities. Some homework should encourage family participation and lighthearted fun. There should be games and puzzles. The quickest child should have some challenge and the slowest some success. All of this is happening in some U.S. fifth grades.

In Japan, only one problem is discussed in a typical forty minute lesson. In the United States, often over ten "problems" are "covered."[2] No wonder children learn so much more math in Japan! The ten problems are either just Drill and Kill, or they are presented so quickly that a thinking being can't catch them as they fly by. Learning mathematics takes time. Recent studies show that the hurried, superficial nature of the U.S. math curriculum is a major cause of our children's abysmal math achievement compared to many other countries.[3]

There appear to be two general versions of Fifth Grade Crisis—feeling overwhelmed and infuriating boredom. The former group are diagnosed (by themselves and others) as underachievers and the latter as overachievers. These two versions, however, may be almost the same. If your child was a bright kindergartener and has not had a devastating accident or illness since, that brightness is probably still intact. If it hasn't been polished, the system is probably more to blame than your child.

> A fifth grader I knew well had several certified disabilities, and the school wanted him to repeat fifth grade. He wasn't quite failing, but they thought he should try again. He was bored, and didn't want to be more bored. The family protested *strongly,* and he was allowed to limp through sixth grade.
>
> When his family moved two years later, his grades shot upward. He now has a plum corporate job, having graduated from a fine university *cum laude* with a double major in mathematics and computer science.

This child was a handsome white male. If your child is not handsome, not white, or not male, you have even more reason to doubt pejorative diagnoses. In most states, school districts get more state money if they "classify" your child. Your child may be smart but bored. Arithmetic may have lost its charm.

I n many other countries, fifth graders study algebra. Not just the top few students, but everyone, except the mentally retarded. Not just the industrial countries, but the developing countries, as soon as universal education reaches fifth grade. Robert Moses of the Algebra Project writes with anger that the "rationing of algebra" in this country "will have devastating long-term effects on the well-being of the entire community."[4] As described on page 243, Mr. Moses successfully teaches algebra to our country's least prepared middle school children, and has discovered that some who had serious trouble with arithmetic excel at algebra.[5, 6]

Too many U.S. fifth graders repeat arithmetic they should have learned in K–4, often with oppressive drill. A child with the misfortune to remember what was taught before is doomed to agonizing boredom. When children are so bored they can't concentrate, they may be labeled slow or naughty, when it's really the curriculum that's slow and naughty.[7]

Algebra is not essential to a good fifth grade program. Statistics, geometry, and logic are certainly defensible, but fifth grade should *not* be review and it *should* focus on solving interesting problems. This is an issue on which the NCTM and its most vocal critics strongly agree; U.S. students are woefully underchallenged in the middle grades.

A high school teacher asked her "lower level" mathematics class if they thought doing many problems was a good way to learn.

"We've always done it that way," they responded. "If we do enough of the problems, eventually we remember how to do them."

"But wouldn't it be better to *understand* how to do the problems?" asked the teacher.

"We've never learned that way."

"Why not?"

Then the class became extremely animated. They claimed they thought their elementary school teachers didn't understand how to do the problems.

"They just did them exactly the way they were in the book. They never taught us anything that wasn't in the book." These students, who had been shunted into the least demanding curriculum, had noticed that their former teachers didn't really know the math. They *realized* this was probably why they had been forced into repetitive drill.

The high school teacher then proposed that she assign only four problems a night, but she would expect all the students to come back with either an answer or a question for each. A month later, she reported "The classes are much more fun. The students come back with questions. Instead of saying, 'I can't do problem 3,' they say, 'I did problem 3 up to this point, and then I couldn't figure out what to do next.' Some of the others will volunteer suggestions as to what to do next. We're having intellectual discussions instead of just one-way explanations."

One must wonder what students might achieve if elementary school teachers all had an adequate math background to stimulate discussion. No matter how well prepared our high school mathematics teachers are, if youngsters are already seriously damaged, we all lose.

On the Frontier of One's Own Knowledge: One mother who was worried about her son's sixth grade mathematics experience visited his teacher. Gently, she asked what his goals were for the remainder of the year.

"We'll do fractions and if there's time, we'll begin geometry."

"Don't you think some pre-algebra skills would be a good idea?" asked the mother. In that system seventh graders officially begin algebra.

"My daughter was in seventh grade last year," responded the teacher. "When she brought her algebra book home, I looked it over. 'You're on your own now!' I told her."

This mother appreciated the teacher's honesty, but her sympathy for her intellectually lively—and bored—son was heightened.

An Unhappy Story: A mathematician observed a fifth grade class and came away furious. Several children offered a better, alternative way to solve a particular problem than the teacher had demonstrated, but she insisted that hers was "the" right way. The children protested that theirs was even better. She replied, "But mine is what the book said to do!"

The kids got mad. "They knew they were right, and they were angry that they were not given proper recognition for their cleverness. I got mad too," said my mathematician friend. "I can't stand being in a class like that. I'm not going back!"

A Happy Story: A mathematics educator "had put up with a lot of very bad mathematics education in my kids' classes." But when her oldest was bringing home 30 exercises a night in fifth grade, she could be patient no longer. She made an appointment with the teacher, and entered the room steaming. The teacher's opening comment caught her off balance.

"I'm so glad you came! I don't know the mathematics I'm supposed to be teaching this year, and nobody ever taught me anything about how to teach mathematics. Can you help me?" The math educator began by explaining the mathematics of the current week's lesson, and spent the rest of the session discussing pedagogy. After that she regularly sent in suggested activities for her son's class. The teacher's approach changed sufficiently for her son to stop hating math lessons. The following year the teacher was given a third grade class and began wholeheartedly implementing modern teaching approaches.

Not every teacher is as open as this one, but it is possible that if the mathematician had spoken to the first teacher, that story too would have had a happy ending. Not every parent is as able as the second, and certainly most parents are not *recognized* as being able to teach math, but you may have more options than you realize. You can offer books and catalogs suggested in Appendix C and maybe can help recruit a consultant.

Another Happy Story: The same math educator became worried about rumors of Drill and Kill in the middle school when her son was in fourth grade. At her urging, the district paid for both sixth grade math teachers to go to a special summer program. Afterward, they completely revamped their approach. Their pupils, including the son of the math educator, enjoyed tackling difficult problems in groups.

She reported, "The teachers are having a ball! And they are both *at least in their late fifties*!"

Don't *ever* think that your child's teachers are to old to learn!

No wonder fifth grade mathematics crises are common in this country! In my professional capacity, I encourage teachers to learn openly from their pupils. There are many elementary school teachers who are comfortable with this and take justifiable pride in their children's superior problem-solving prowess. However, it isn't prudent for a parent to assume that a child's teacher will welcome being taught on the job, and it is wise to warn your child to be wary of the risks.

If your child faces one or more symptoms of Fifth Grade Crisis, ignoring the situation is *not* recommended. Such troubles are *serious*. Preadolescence is a delicate time, and bending the twig too low can impair the subsequent growth of the tree.

Helping the Teacher

If you are comfortable having a chat with your child's teacher, you're lucky. Putting your heads together may point the way to success for your child.

One sixth grade teacher suggests that instead of directly confronting your child's teacher, you should ask the principal if that teacher is routinely sent to math conferences and workshops. If not, you should emphasize the importance of permitting teachers to learn mathematics. Teachers with plenty of ideas about how to make math exciting rarely teach boring lessons.

If the principal and you conclude the teacher is not getting enough support, the two of you can discuss how to *help* the teacher. Is there a math specialist who can visit the class regularly? Does the teacher have a copy of the NCTM *Standards*? Has the principal encouraged teachers to read math journals, use cooperative learning, and discuss alternative solutions? Are appropriate books and teaching aids available? Are manipulatives available? Are there regular in-service workshops about mathematics and resourceful ways to teach it? Does the school lend out inspiring and stimulating videotapes like those of Marilyn Burns[8] and of Constance Kamii?[9] What are the roles of PTA's and professional committees in implementing change? Are there other teachers who need similar support? These questions may help the principal focus on the need for significant mathematical change. Perhaps the PTA as a group, not you as an individual, should ask the questions.

Some teachers may resent your approaching the principal first; you have to make a delicate decision. Some educators believe that it is *unethical* to approach the principal without first discussing an issue with your child's teacher. Such people believe the proper sequence in which to express any concern is teacher, principal, district mathematics supervisor, curriculum director, superintendent, and school board; parents who "go over the head" of any of these without warning deserve whatever penalty fate bestows upon them.

If you are comfortable talking first to the teacher, you might raise similar questions about the teacher's need for math enrichment and offer to visit the principal or curriculum director on her or his behalf. The teacher may decide your pleading is better than complaining, and so will agree to this offer. If not, then you have to make another delicate decision. Remember that the teacher may feel both relief and anger about your offer where only one is evident.

It takes time to change one's teaching techniques. Marilyn Burns advocates implementing one new teaching technique a month. This is sufficiently nonthreatening that many teachers are willing to try it, and at the end of one school year, the teacher has ten new approaches, enough to be a drastic shift. One change a month may not seem wonderful if your child already has the teacher, but causing some change (via you) may foster patience and relieve your child's sense of powerlessness. It may also keep your child from developing a hatred of mathematics and/or school.

I believe that one of the most cost-effective drug prevention programs would be to empower teachers in grades 5–8 to provide exciting, ego-boosting thrills in mathematics classes. Most teachers are willing, but they can't teach themselves mathematics alone, especially not while teaching full time. Supporting the continuing education of middle school math teachers could have great benefits for our society—not just mathematically.

What can kind but mathematically underprepared teachers do to avoid squelching the enthusiasm of an intelligent, inquisitive child? They could offer "canned" activities to challenge all their children while including even the slowest. The best teachers do this phenomenally well. Perhaps they can let your child work on appropriate material in the back of the room. Computer programs sometimes provide really sophisticated challenges (see Appendix B). Your child might tutor others, deepening mathematical understanding while developing social skills. Or perhaps the restless souls will play games like chess, go, and owari, which challenge the finest intellect and are accessible to children.

It's important to support teachers who occasionally put your "gifted" child in the back of the room with a book, computer, game, or another child to tutor. These are worthwhile activities. A teacher's job is not always to be performing, but to stimulate intellectual growth. Tutoring, reading and playing games should *not* be viewed as escapes for the teacher, but as realistic responses to your fine parenting.

Helping the Relationship

If a personality conflict between the teacher and your child is seriously interfering with academic progress, you will certainly want mediation. You are the first mediator, but not the best one because you are biased on behalf of your child. At least, we hope you are! Remember that your child may not be at her or his most lovable in math class, so try to have compassion for both the teacher and the child.

Whether the teacher perceives your child as too stupid or too smart (in fifth grade, as we noted above, the results can be the same), the teacher may see no options for responding to nonconformity. Perhaps someone else can suggest new ways of expanding math horizons. A district math supervisor may be helpful.

Sometimes tincture of time remedies a child's rebellion against a teacher's style.

A Strict Teacher's Legacy: The late Pat Podesta was a strict middle school English teacher. When my daughter first had her in fifth grade, she resented Ms. Podesta's serious demeanor and critical comments. However, when her younger brother rebelled against Ms. Podesta, she assured him that he would grow to love his demanding new teacher. He doubted it, but he too left middle school very fond of Pat Podesta.

Sometimes there is no animosity, but just something missing in the teacher-student relationship.

I Don't *See* . . . : My husband was failing Algebra 2 until his teacher got sick. The substitute watched him for a few days, and then asked him to come in for help after school.

"Why aren't you doing better?" asked the substitute. I can tell you are much smarter than your grades show."

Fred had been wondering the same thing. "Somehow I just don't *see* . . ." He hesitated.

"See! You learn mathematics visually! Let me show you algebra in pictures!"

He did. Now Fred has over forty published research papers in electrical engineering, using sophisticated advanced mathematics.

One can only wonder how different Fred's life might be if his original teacher hadn't gotten sick. One can also wonder how many children are needlessly alienated from mathematics because geometry is so disconnected from numbers and algebra in this culture (earlier chapters give examples of connections).

Sometimes teacher-student troubles merit anonymous intervention.

Anonymous Gifts: My daughter once had a mathematically competent teacher who thought sexist behavior was funny. I was reluctant to confront him because of possible reprisals on my child.

Finally we hit on the idea of buying him, anonymously, a membership in the Association for Women in Mathematics! Every other month he received an AWM newsletter with articles about remedying sexism in the classroom and relating achievements of women in mathematics.

He was a quick learner. Almost immediately, my daughter reported progress, and a couple of years later my spies in his class reported that he had completely reformed. Now he spouted the AWM party lines instead of sexist "jokes"! We still wonder whether he figured out who paid for his AWM membership.[10]

There are other possible gifts you can buy an errant teacher. Consider the lists in Appendices A and B. Membership in the NAACP (National Association for the Advancement of Colored People) brings the excellent magazine *Crisis*[11], which might have a similar impact on a racist teacher that the AWM newsletter did in my story. The magazine *Teaching Tolerance* is free to educators and addresses a variety of intergroup issues.[12]

A membership in the National Council of Teachers of Mathematics[13] can bring a subscription to *Teaching Children Mathematics*. Monthly issues of this fine magazine will probably affect a teacher's approach. The cost is modest compared to your child's soul, or even the cost of a private school, or even the cost of tutors and/or therapy.

Helping Your Child

How do we help our children cope with the evil of this world? This question pervades parenthood. Just coping can be very difficult.

If your child's situation is lamentable, let your child lament to you. People under stress need to have their stories heard, and your child is under serious stress. Don't underestimate the Fifth Grade Crisis. Multiplied by millions, it is destroying our culture and economy.

An expensive, prestigious study about how parents can improve their children's achievement concluded that "talking regularly about current school experiences" is highly significant.[14] "Current" implies that "talking" means conversing. A familiar theme. If your child perceives you as a listening parent, adolescence will be far easier. Now is the time to firm up lines of communication. Tell your child about some of your own troubles and successes in adapting to difficult and/or new situations. Remembering your own childhood troubles may be interesting even though your child still doubts that you were ever a child. If you don't overdo it (your child is *not* your therapist!), your child may enjoy being treated like a confidant.

If your child has already succumbed to math anxiety, Sheila Tobias' classic *Math Anxiety*[15] may help. She emphasizes Hugh Rossi's observation mentioned in Chapter 3 (see p. 31): everyone becomes confused while doing real mathematics. Don't take it personally. Don't expect to do math quickly and easily. Real math is hard, but "hard" need not mean "unpleasant."[16] Two more recent books with a similar theme are Claudia Zaslavsky's *Fear of Math*[17] and Marilyn Frankenstein's *Relearning Mathematics*.[18] All three are written for adults, but they may help you help your child. Reading passages aloud may give your child ideas. Actually, many middle school folks enjoy adult books. Why not?

The politics of coping are muddy, and not easy to teach. Remember the story in the previous chapter about my mother suggesting it might be better not to tell Miss Mitchell about negative numbers. Your child may now be old enough to share some secrets. Reading this chapter together may be appropriate.

Even if your consultations at school seem ineffective, they may bring hope and provide emotional support for your child. Relay my sympathy. If boring homework is a factor, you might waive normally sensible household rules against watching TV or listening to distracting music (as suggested in the previous chapter). The Third International Mathematics and Science Study indicated that U.S. students are assigned more homework than children in countries with higher math achievement; the implication is that the homework is not well chosen.[19] Although parents teach their children that sometimes we must all do things we don't want to do, we don't want children to acquiesce unquestioningly to destructive rules. If appropriate, remind your child that such administrators and teachers are merely doing what they were taught. Understanding sometimes generates tolerance.

Other leaders in your child's life—for example, a religious teacher, a scout leader, or a Tae Kwan Do instructor—may help your child be tolerant in an imperfect world while working for change.

> **A Nice Teacher:** I asked a seventh grader if algebra was boring. "No actually, it isn't. For the first time, I have a teacher who knows more math than I do. I'm learning from her. It's sorta nice."
>
> "How did you get along with your sixth grade teacher?"
>
> "Not too badly. She didn't know how little math she knew. We didn't tell her."
>
> "She did the best she could under the circumstances," said his mother, who is a mathematician. "She was a nice person."

Obviously, this son and his mother had had significant conversations about his math education. He was far behind where he might have been if our schools challenged good students, but he had learned to be patient with grown-ups. If it seems merited, praise your child for being kind to the teacher.

Learning mathematics secretly may appeal.

> **At the Ironing Board:** When I was ten, mothers regularly spent time ironing, a time when they were available for conversations. One such time, I asked mine what algebra was. She suggested that I look up "algebra" in our encyclopedia. While she ironed, we went through algebra as explained in the *Britannica Junior*. Somehow, learning algebra made arithmetic seem more palatable, although it made ninth grade algebra somewhat boring. But, fortunately, the late Belle Kearny supplemented the curriculum and inspired me.

To avoid the risks of teaching traditional math before your child meets it in school, you can stimulate math through puzzles. Books of zany puzzles are available in your library, bookstore, and the catalogs in Appendix C. Fifth graders enjoy old classics such as, "A bear walked one mile south, one mile east, one mile north, and arrived at her starting point. What color is the bear?"[20]

Fifth graders are not too old for games such as those described in Chapter 9. The ancient intellectual games of Europe, Asia, and Africa (chess, go, and owari) can be as much fun outside school as in. If your child becomes a fan of one of these, mathematical growth is assured. Perhaps you and your child can join or begin a club. Tournaments in these games (and the modern game of Equations) can provide the excitement often reserved in this country for sports.

Seize every opportunity to take your children to science and math displays in museums. If you can take along friends of your children, so much the better. Do you know someone working in math or science who would tell a group of middle school children about careers? Combining social and intellectual escapades improves both.

Not Believing Tests

Labels limit. If your child is officially labeled, the stigma may not be easy to shake. But try not to let early pigeonholes become a life sentence.

Certainly, if your child has a genuine handicap *and expert teaching is provided,* that is fine. It's worth the label. But if your school claims that your child "fell behind grade level" in the middle grades (4–8), make your own independent observations. Remember the power of self-fulfilling prophecies, and read the next five chapters carefully.

When children are treated like robots, they don't like it. Low expectations can wreak havoc. When a child stops trying, tests scores don't reflect either ability or achievement. Perhaps your child thinks as clearly as you do. Why comply with stupid rules and tests? I write with passion, because my own child almost fell into the pit.

> **Certified Incorrigible:** My firstborn had a teacher that gave even more Drill and Kill to the smart kids than the average ones. Lori said she did not deserve to be punished for being smart and refused to do the assignments. The teacher recommended she be evaluated. Five "experts" studied her and then summoned me.
>
> "She is incorrigible and will never amount to anything," they reported sternly. I said I didn't believe it. "It's your fault!" the experts continued. Obviously, I was incorrigible too. Patiently, they explained to me, as one explains to a naughty child, that I had ruined my child. It was an extremely unpleasant experience.

Fortunately for Lori and me, my parents had both the good sense and the money to help me immediately transfer her to a private school and move out of town that summer. When we came to Montclair, Lori got Elizabeth Ivansheck Smith for math. The first night's homework included a varied set of problems. My child, who had refused to do last year's boring assignments, zoomed through this set and was stuck at the last problem.

She brought up her last homework question at dinner. "Prove half of thirteen is eight." Our family is not used to being stumped on math homework problems, and we put our collective intelligence into solving this one. The next day, Lori was the only child with the answer—a great boon to both her ego and her sense of adventure.[21]

Elizabeth Ivansheck Smith, Pat Podesta, and their colleagues turned around Lori's life.

Many parents don't have the options I had, but their children aren't incorrigible either. Today, parental choice within public schools seems necessary and is a growing trend. With luck, an "incorrigible" child can get a fresh start within the same system, preferably in the same school.

"Incorrigible" children need love and support, like everybody else. Perhaps they are children who see the folly of our society better than others and refuse to be sucked into structured silliness. My experience as a licensed emergency foster parent of other families' "incorrigible" teenagers (when my own kids were in high school and no longer considered troublesome) supports this suspicion. Drill and Kill in preadolescence is causing vastly more tragedy in our country than is generally acknowledged—both to our children and to their parents.

Misdiagnosis, low expectations, and their accompanying rebelliousness are even more rampant among minorities, poor families, and those with less education and less money than in my favored social milieu—where they are entirely too rampant. Our country desperately needs to provide intellectually stimulating and compassionate education for *all* children. As the recent ACORN "Secret Apartheid" report asks, "Why are the innovative approaches used in gifted programs not used in all of our schools?"[22] Meanwhile, parents alert enough to be reading this book need to struggle both for their own children and for systemic change.

Most ten-year-olds—and probably *your* ten-year-old—perceive the evil of passing off repetitive exercises as "education." Most ten-year-olds have a strong sense of right and wrong, but they haven't yet experienced enough of life's quixotic cruelties to realize how easy it is to err and how divine to forgive. They just know that forcing children to waste the precious gift of life at merciless drill is *wrong*. It is; in that, they are right.

WHY SO MANY CHILDREN ARE DAMAGED

Covenant Dishonored: Three Theories of Educational Reform

15

We want a class of persons to have a liberal education, and we want another class of persons, a very much larger class of necessity in every society, to forego the privilege of a liberal education and fit themselves to perform specific difficult manual tasks.

—*U.S. President Woodrow Wilson, president of Princeton University.*[1]

Why would any politician wish to confront an informed citizenry that could read the federal budget . . . ? Woodrow Wilson's generation took it upon themselves to rig the curricula in a way that discouraged the habits of skepticism or dissent. . . .

If many of our public schools resemble penal institutions, . . . then I cannot help but think that the result is neither an accident nor a mistake. We make it as difficult as possible for our children to learn anything other than their proper place in the economic order because we fear the power of untrammeled thought. . . .

—*Lewis H. Lapham*[2]

The dream of an educational system that honors its covenant with families to provide an excellent education for *all* children is still a dream. There have always been many superb teachers in the United States. There still are. However, the *system* too often produces serious disillusionment.

Lapham argues convincingly in his article quoted above[3] that our traditional system was never really designed to educate the working class, girls, or descendents of slaves. United

States public schools have disappointed many parents who believed that the schools would cultivate their children's intellects. Those labeled "gifted" might get the best public education in the world. Otherwise, parents who trusted were often betrayed. As mentioned at the beginning of Chapter 1, even our best do poorly on international standardized tests.

I am deeply saddened by what passes for education in many U.S. public schools, not just in disadvantaged areas but even in some of the "best" districts. We have gotten off track. Most of us want to return to the pursuit of excellence.

The custom of pitting self-identified elitists against self-identified egalitarians is damaging both concerns. How can you care about the less privileged without insisting that they be allowed to achieve at high levels? How can you really love knowledge without wanting to share it with everyone? Educational leaders like Marva Collins, Robert Moses, and Seymour Spiegel have proved that excellence and equity are compatible. They have taken in "the least of these" and produced outstanding professional leaders. I believe that true equity and true excellence are not only compatible, but inseparable.

Nevertheless, until recently, even the *rhetoric* of educational policy neglected many children, as shown in President Wilson's quote. Until recently, most young adults could get jobs on farms, in other people's homes, or in factories. Intellectual achievement was not needed for a satisfactory niche in the economy. Today, it's different.

Jobs are changing rapidly and disappearing. Without a serious intellectual education, U.S. children are not being prepared to compete in a worldwide economy based on sophisticated mathematics. They may grow up with few alternatives to welfare or crime.

While educational rhetoric has changed, actual change within the schools is excruciatingly slow. We need to look thoughtfully at what is happening in our schools and not be duped by rhetoric.

What Is an Excellent School?

How would a parent recognize an excellent school if it were around the corner? An excellent school teaches:

- a love of learning
- many facts
- an ability to solve new problems and learn new topics

These three criteria are the underpinning of three current approaches to educational reform (described in this chapter) whose adherents debate heatedly. The heat has gone beyond constructive warming. Our children need all three: a love of learning, knowledge of many facts, and a growing ability to solve problems and learn. The balance is arguable, but none should be ignored.

How Does a School Achieve Excellence?

If this question had a simple answer, fixing our schools would take only commitment and money. Alas, it isn't that straightforward. There have been many studies, and some contradict others. However, four conclusions seem to be common to all the studies:

1. Parents are involved.
2. High achievement is expected of students.
3. Principals are strong. They have freedom to follow their vision and they attract or convince teachers and parents who share that vision.
4. Teachers have the professional freedom to plan, in consultation with each other, lessons and activities that meet the needs of their particular students.

Let us examine each.

1. Your children need your support and concern for their education, right up through high school. Most of the time, you will be collaborating with their teachers. Even if their teachers are fabulous, your participation is indispensable.

2. Low expectations are the norm. "Only in the United States do people believe that learning mathematics depends on special ability. In other countries, students, parents, and teachers all expect that most students can master mathematics if only they work hard enough."[4]

Tests that "measure ability" are sometimes used to place children *as young as kindergarten* in tracks. Catching up becomes very difficult, if not impossible.

3. State regulations, although well-intentioned, are no substitute for a strong, talented local leader who knows the community and the children. Indeed, legislated restrictions can seriously hamper effective leadership. In a country with many types of communities accustomed to local control, regulation and standardized testing can do terrible harm. Data from 500 randomly selected high schools nationwide indicate that the most important factor in making a school effective is autonomy.[5]

On the other hand, we need higher mathematics requirements as a *pre*requisite to teaching. All our states' math requirements for prospective elementary school teachers are far lower than those of other countries, and 24 states had none whatsoever in 1995.[6]

4. Teachers must believe in what they are expected to do. They need opportunities to collaborate and freedom to implement what they commit to do. Most teachers are caring and capable. Too many have been crushed under an oppressive, uninspiring system.

Alternative Approaches to Education

There are many fine approaches to education, but the adults in any one school must have some common beliefs. If they squabble too much, the children lose. Choices can be provided within the public schools.

Twenty years ago, my hometown of Montclair introduced voluntary bussing to magnet schools to racially desegregate the schools. This worked, and many advantages soon became apparent having nothing to do with race. As schools focused on more important common traits than race, everyone became happier. All children benefited.

The titles of Montclair's magnets now include, "Gifted and Talented" (focusing on arts and music), "International" (where children from thirty-five different countries with twenty native languages have formed their own "United Nations"), "Math and Science," "Family Centered," "Montessori," and "Technology." They adopt good ideas from each other and aren't as specialized as their names indicate, but their special emphasis enables each to feel pride in being "the best." Thus the American competitive instinct is channeled constructively. As I've watched the magnets thrive for two decades, I've become convinced there are many ways a cohesive team of parents and professionals can evolve a fine learning environment. The best setting for any one child depends on many factors. One advantage of choice within the public system is that if a school becomes unsuitable for a child, a new start is possible without moving from town.

Meanwhile, other districts (and even the entire state of Minnesota) began offering choice within the public school system. One of the most spectacular successes is District Four in New York City. In fourteen years, the reported percentage of its students reading at grade level jumped from 16% to 63%. A district that once failed essentially all of its students began producing an impressive number of "winners."[7]

Myron Lieberman[8] argues that it wasn't choice itself but the creative approaches that choice allowed innovative educators to implement that caused the "miracle." Nobody seems to deny this. Unfortunately, however, choice seems to be needed currently to give dedicated educators the freedom to counter the inertia of our system. "The bureaucracy can always persevere; since it is used to doing nothing it can hold on and wait an innovator out," growls Seymour Fliegel, a major player in New York City's District Four turnabout.[9] If the entire system becomes more effective some day, we may outgrow our need for choice. Meanwhile, some research seems to indicate that when other variables are held constant, children in magnet schools score higher on standardized tests.[10] Most of these do not test *creative* mathematics, but they are the best scales we currently have for comparing students' achievement in different learning environments.

Choice works, obviously, only if there are good choices available; just as obviously, choice "should not be confused with comprehensive school reform—which is both more complex and more expensive."[11] And almost everyone who studies U.S. schools agrees that they need significant reform. A prestigious report compiled by leading educators and politicians led by North Carolina Governor James B. Hunt, Jr., advocates a variety of reforms consistent with this book without mentioning the issue of choice.[12] Obviously, choice will not improve the overall national picture unless all children have good choices.

Even when good choices are available, they help only those who can make them. A study by a community group in New York City indicates that white parents are generally treated better and offered more options than black parents when they try to get information about the best school for their prospective kindergartner.[13] If "choice" becomes an excuse for *denying* poor children options, as it sometimes has been in the past, then we may need to abandon it. Decades hence, "choice" may be seen as the diversionary tactic that the 1960s effort for integration is now regarded by some. "[School] integration for black people has been almost a cruel hoax," bemoans the president of the Institute for Independent Education,[14] which helps blacks, Latinos, and Native Americans create and nourish their own alternative schooling. Currently, however, choice within the public school system for both teachers and parents seems to me to be a promising tool for jogging the inertia of our sluggish system. Not having educational choice is, in effect, like having arranged marriages with impersonal machines deciding a youngster's fate.

National Debate: Three Theories of Reform

Nationally, there seem to be three major theories about reforming education in the 1990s. Of course, there are actually as many theories as there are people concerned about education. While there are gray areas among the three described here, there do seem to be three major themes.[15] There are no generally accepted names for these approaches, so I will make some up. I suggest two names for each, one that the followers of the theme might choose and another that the foes do use.

1. **The Self-Esteem or Feel-Good Theory:** Supporters of this theory assert that building children's self-esteem is basic to education because it empowers children to learn. Methods of teaching seem very important, and content less so. If a child enjoys learning, the subject matter will follow. The self-esteem group talks a great deal about equity issues, noting that faltering student egos have been a major cause of low achievement. Currently, a major goal is to "detrack" schools.

2. **The Accountability or Authoritarian Theory:** This group asserts that there is "basic" knowledge that all children should learn and that we can measure whether a particular child has learned it. They believe that central authorities can set standards, and individual teachers and schools could be held "accountable" to those standards. They believe that

state or national tests can measure achievement and that those who don't measure up should be "failed."

3. **The Relational or Teacher-Dependent Theory:** Its proponents talk about the relationships

- ▸ within a subject (what is the relationship of three, four, and twelve in multiplication?),
- ▸ between a student and subject (does $3 \times 4 = 12$ make sense to the child or is it mere memorization?), and
- ▸ between student and teacher (who may be a parent).

The subject matter preparation of paid teachers seems crucial, along with their attitude, class size, and working conditions.

The first theory dominates our schools of education and the second our media and politicians. From my vantage point in the third, it seems that advocates of the first two confuse each other with the third, so their potshots at each other land on teachers and other intellectuals. Too little attention is being paid to how knowledge and ideas are generated and transmitted.

PROBLEMS WITH THE SELF-ESTEEM THEORY

Too often the feel-good philosophy accomplishes little or nothing. Just as the *pursuit* of happiness may be one of the *least* reliable paths to happiness, too much direct emphasis on self-esteem may be counterproductive. The actual learning of mathematics, in contrast, can raise self-esteem remarkably.[16]

Indeed, the best teaching may be so subtle that it is hardly perceived as teaching. It's good for everyone's self-esteem. Assaults on a child's ego can indeed be very damaging to educational success, but disrespect is not the *only* cause of educational disappointment. Lack of access to intellectual ideas looms large among the obstacles.

Children who are not taught major mathematical concepts rarely learn them. When I asked 75 black mathematicians of New Jersey what can be done to bring more blacks into mathematics, the leading answer was to teach all American children mathematics better in elementary school. "The way it is now, if you don't learn math at home, you don't learn it at all, so any ethnic group that is underrepresented in mathematics will remain so."[17]

Thus, we can view the current custom of forcing prospective elementary school teachers to take many courses in educational theory and virtually none in mathematics as racist.[18] Such policies prevent access to math and science for ethnic groups whose families do not teach them math at home. Since math and science provide paths to satisfying

Example from History

Carl Friedrich Gauss (1777–1855) has been called the greatest mathematician of all time. He was the son of a stone cutter in Prussia (now eastern Germany) when public education was becoming available for sons of the working class.

One day seven-year-old Carl's teacher was exasperated with his young scholars and told them to add all the numbers from one to 100. He thought that would keep them busy for a long time, but little Carl did the sum in a minute. The teacher brought him to the attention of the nearby nobleman, and Gauss's life changed.

Let's consider Gauss's problem from three points of view. A self-esteem advocate might modify the problem so that they add only the numbers from one to ten. An accountability type might insist they grind away at the long problem, one boring number at a time. A relational teacher might help a more normal child see the problem as little Gauss did.

Teacher: You seem unhappy with this problem. Could you do an easier problem?

Student: I'd rather add just the numbers from one to ten!

Teacher: Why could you do that more easily?

Student: It wouldn't take so long because there aren't so many numbers.

Teacher: How can you make it even easier? Which two numbers would you add first?

Student: I might try the biggest two. Or the smallest two. Or maybe one of each.

Teacher: What happens if you take one of each?

Student: One plus ten is eleven. . . . There are other elevens!

Teacher: How many elevens? (The teacher holds up fingers as the student lists sums.)

Student: $1 + 10, 2 + 9, 3 + 8, 4 + 7$, and $5 + 6$, so there are five elevens—55 in all.

Teacher: Good! You can do the easier problem in an easy way. That is promising. Can you do the harder problem in a similar way?

Student: The largest and smallest numbers are 100 and 1. That makes 101. How many of them are there? That's a lot harder than before.

Teacher: How might you find out?

Student: Well, there is $1 + 100, 2 + 99, 3 + 98$. . . . I guess I can skip. What do I skip to? Let's see. $50 + 51 = 101$ so that's the last of the pairs. There must be 50 pairs.

Teacher: If there are 50 pairs, each adding to 101, how can you find the whole sum?

Student: I have to multiply 50 by 101, which is multiplying 50 by 100 and adding it to 50 times 1 or just plain 50. Now, 50 times 100 is 5,000. So the whole sum is 5,050.

Pretty good for a seven-year-old to do alone! But my husband loves to guide kids through this story in social settings. The parents are surprised, but most children seem to enjoy it.

first-generation professional careers, the long-term consequences of depriving teachers knowledge of these subjects is worst for previously excluded groups.

If parents who are not math-confident themselves can teach the math they do know and learn *with* their children, their children are likely to develop enough math ability to succeed when they do meet math-prepared teachers. An elementary school teacher, however, will have a more difficult time learning math while teaching it because there are so many more children. With help, teachers, like parents, can (almost?) catch up to their pupils, but they do need significant help.

It is true that subject matter knowledge alone is not enough to make a good teacher, but it is essential. You can't teach or learn with a closed mind, but it can't be too open either. There is a balance between cluttering a mind with so much secondhand knowledge that it can't "see the woods for the trees" and leaving it so empty that nothing relates to anything else.

Furthermore, the extent to which taking courses in education actually improves teachers' teaching abilities is not clear. There is no single "correct" way to teach. Subject matter courses taught by faculty who model effective teaching techniques also provide pedagogical choices to prospective teachers. U.S. teachers need to take math and science courses comparable to those required of teachers in other countries. It is impossible to teach information you don't know.

PROBLEMS WITH THE ACCOUNTABILITY THEORY

The authoritarian approach has so many problems that I have devoted four chapters to it: two about Drill and Kill teaching and two about testing. Who decides accountability? Using what method?

One of accountability's most articulate proponents, Charles Sykes, includes a devastating chapter about testing companies in his book *Dumbing Down Our Kids*.[19] One must wonder who he expects to enforce accountability, if testing companies are as he describes. The widely acknowledged horrors of U.S. testing are major hurdles for the accountability camp. One of the most comprehensive books about U.S. testing says in its introduction, "If there is a single lesson to be learned from our review of the testing marketplace, it is simply that we must be more careful than we have in the past about relying too greatly on imperfect test instruments arising from the highly fractured market surrounding testing."[20] The book warns that when tests are mandated by state governments, responsibility is blurred,[21] too often leading to "false and misleading inferences from test results,"[22] especially if the test was written hastily for money.[23]

Is there a way to establish an accountability system better than testing companies? New Jersey has experimented with a plausible system. Leading state educators and mathematicians have collaborated to create two examinations that test what mathematicians consider important in ways that educators deem reliable.[24] One is called the High School Proficiency Test (HSPT) and is required for state-sanctioned graduation. It is given to eleventh graders so they have a year to study and retake it without delaying graduation. The other, the Early Warning Test (EWT), is given to eighth graders to see if they are making progress toward passing the HSPT. Both exams test ideas and skills taught in good elementary schools. The professionals who monitor them do so not for money but out of social conscience. They are fine tests.

Sounds good? Not when you read Jonathan Kozol's account of their effect in *Savage Inequalities*! One urban New Jersey principal (Ruthie Green-Brown, principal of Camden High) observes, "If they first had given Head Start to our children *and* pre-kindergarten, *and* materials *and* classes of 15 or 18 children in the elementary grades, *and* computers *and* attractive buildings *and* enough books and supplies *and* teacher salaries sufficient to compete with the suburban schools, and then come in a few years later with their tests and test-demands, it might have been fair play."[25] Instead, about 80% of the urban youngsters fail.

Then they are taught to the tests by rote with no conceptual framework. High school has become strictly remedial. The state threatens to withdraw funds if the students don't pass. "From September to May . . . instruction is exclusively test preparation. Then if we are lucky, we have two months left . . . to teach some subject matter," laments a Camden High teacher.[26]

"Already in ninth grade the youngsters are saying, 'If I have to do this all again, I'm leaving,'" continues Green-Brown.[27] Most fail again that year. Many drop out then or in tenth grade. The consequences of our excellent, carefully conceived, state tests have doomed these teens to the streets.[28]

This story points to another major problem with "accountability": It begets boring, time-wasting drill as teachers frantically try to put facts into youngsters' short-term memory. One can claim it is not the tests themselves that *cause* the drill. But pressured administrators and teachers do, in fact, respond to authoritarian tests with the type of drill that kills collaboration, joy, hope, and ability itself. Their haste makes not just waste, but destruction of future options.

Accountability doesn't teach. In its best form it tells society how badly we are doing. In its worst, it takes much needed resources (time and money[29]) from teaching and helping elementary school teachers. Drill and Kill imposed in response to authoritarian tests prevents children from learning.

PROBLEMS WITH THE RELATIONAL THEORY

The major problem of relational education is that teachers must know the subject matter reasonably well and also be familiar with a variety of ways to teach it. Teachers must set the direction and "hear" what the students really mean when they ask questions. Teachers plant the seeds of knowledge, cultivate the students' minds, and listen with caring, perceptive ears. The arts of questioning and problem presenting are crucial. In a country where twenty-four of the states have no math requirements at all for prospective elementary school teachers and the others have very few by international standards, depending on teachers is problematic, as noted before.

School structures present other problems. Well-prepared teachers ready to teach mathematics imaginatively and joyfully need and deserve much more professional freedom than U.S. teachers currently have. They need time to confer and plan with each other. They need more support in discipline and other nonprofessional activities. Those serving the most needy children need much more of what money can buy.

Relational teaching has no clear podium. We clamor to be heard, but thus far have reached neither the media nor the teacher-preparation programs, except marginally. We need many more writers, educators, parents, teachers, and politicians championing capable teachers' sharing mathematics with their own individual students. Studies repeatedly indicate that the students of such teachers feel good about mathematics and achieve adequate scores on reasonable tests.

The National Council of Teachers of Mathematics (NCTM) has used what power it has to promulgate three sets of *"Standards"* to encourage learning environments that vary with the teacher, the children, and the ideas (see Chapters 11 and 21). It recommends a variety of teaching techniques, including collaborative learning, writing to learn, games and manipulatives, computer interaction, and integration with other subjects. The choice depends upon the needs of the students, the teacher's personality and preparation, and the setting. All of this is consistent with the other two theories, but the ability of teachers to make informed choices is basic to the relational approach.

The Curriculum

Curriculum seems to generate enormous debate. In mathematics, unlike history or literature, the subject matter itself isn't very controversial. Even across continents, where there has been practically no collaboration while devising math programs, researchers have discovered that over 80% of the math content in Japanese elementary school texts is also in U.S. elementary math texts, and vice versa.[30] The recently released TIMSS results indicate that almost all countries have a similar math curriculum. Math content is not a real problem; there must be some other agenda to sustain such debate.

Who decides what will happen in our schools? Some people want the federal government to "fix it" by issuing a fiat. There are two problems: (1) the folks in Washington don't have the wisdom to know what is best for everyone else, and (2) Americans like their local freedom. I do too. Educating people about their options is the best hope for education (see Chapters 20 and 21).

In a country where there is no national or state curriculum, textbooks wield enormous power. Unless local educators are diligent and imaginative, the texts' perceived content may determine (or at least circumscribe) the curriculum. Although their math content is not greatly different from their slim Japanese counterparts, their large size is daunting. Teachers tend to pick and choose what they will teach, leaving mortal gaps in their students' backgrounds.

"American textbooks . . . are thick, hardcover volumes [with] colorful illustrations, photographs, drawings, or figures on nearly every page." "One gets the impression . . . the teacher may omit whole chapters." [31]

Japanese texts are lighter and much smaller than U.S. texts. They are paperback.[32] "Textbooks that contain short lessons, a limited number of practice problems, and practically no ancillary material make it possible for the class to cover every detail contained in every textbook." [33] Japanese children are actually taught all the material for each grade; their teachers are not intimidated by the overwhelming size of the books. They can see what is essential.

Another effect of U.S. textbooks is the way they discourage relational education. Not only the teachers' initiative but also the students' creativity is dwarfed by the all-encompassing book.

"If all the steps in the application of a concept or skill are presented, as in American textbooks, children must simply follow the argument, without encountering gaps that they must figure out for themselves. Since one goal of Asian teachers is to have children learn that there are many different methods for solving problems, presenting a single method in such full detail would limit the likelihood that children would come up with alternative solutions. A common technique used by Asian teachers in mathematics classes is to have children present as many different solutions to a problem as possible, and then to have the class discuss which methods are most efficient, and why." [34]

This quote may reflect *the* major reason for the drastic difference in achievement levels between the U.S. and other countries. Parents need to seize every opportunity to help their children cherish alternative methods to problem solving before schools squelch this natural urge. Cheerfully comparing different solutions is a valuable way to relate to other people. The impact on adult habits is clear. Children who learn to welcome many proposed solutions and to discuss which is best are better prepared to accept and discuss corporate and community decisions as adults.

HOW ARE TEXTBOOKS WRITTEN AND CHOSEN?

Who benefits from our large, expensive texts? The United States has a $4 billion textbook industry.[35] The people employed in these companies work hard. They generate new, attractive packages in consultation with people actively working with children. Indeed, many people on the professional staffs of textbook publishers are former teachers.

Still, it is not easy to buck trends. It can be difficult even to decipher the rhetoric to find out what the trends actually are. Making change is frustrating. The NCTM *Standards* emphasize problem solving, looking at the big picture, and urging children to think creatively. The unified collaboration in promulgating these ideas has been inspiring, but no community is without its misunderstandings.

One person who allegedly bucked NCTM trends is John Saxon[36] (1923–1996), but that may not be completely true. His company now distributes charts indicating how much his texts, lesson by lesson, support the NCTM *Standards*. There are many black dots, indicating that a particular lesson teaches a particular Standard.

Saxon first wrote his Algebra 1 text around 1980 without a traditional background for access to mainline publishers. He took a second mortgage on his home to self-publish. Now his own company has sold millions of his math texts from kindergarten through calculus.

Obviously, many teachers like his books. The before-and-after standardized test scores of students who have used his texts are impressive. Even more convincing, "his" students voluntarily take many more serious high school math courses than their predecessors. Yet the Saxon books have had limited success with state-level adoptions, and have not yet been adopted in the two large, trendsetting states.

Where's the rub? It is a very sad personality and political conflict. The Saxon books do present big ideas and stimulate creative thinking, but neither Saxon's advertising nor his detractors seem to recognize this. Neither side emphasizes how, by presenting each lesson individually, his texts eliminate the need for teachers to spend time planning lessons and devising daily homework. As in Japanese texts, the material is all there, but not too much. Neither side mentions how well his lessons are written, so wobbly teachers can reinforce their backgrounds as they go along, and students have a fighting chance to teach themselves no matter what their teachers' ability.

Instead, both sides debate his exercise sets. Each assignment has no more than five problems from one day's lesson and plenty of review from a variety of previous lessons. The hardest problems generally come first, tapering off to easier ones as the student gets tired. Despite accusations to the contrary, the exercise sets are definitely not Drill and Kill. Saxon's assignments have far more variety than those in most texts. Many of the individual exercises are lively too.

A downside to Saxon's material might be that its detail could become tiresome to a creative teacher as the years go by. Some schools rotate grades among teachers to minimize Saxon burnout. However, a system with overworked teachers (especially if some are also underprepared) could find the Saxon books just the steady crutch they and their students need.

John Saxon has stirred political conflict by venting his considerable anger at how badly the current system has failed, needlessly, in his opinion. His rage has alienated many members of the NCTM. It is hard to see the value in someone else's work while he or she attacks yours. I believe that U.S. math could take a great leap forward if the Saxon movement and the NCTM would collaborate. They have much in common and we need as many voices as possible speaking out for real reform, which both groups want.[37]

The Saxon story illustrates problems with all three current theories of education. Concentrating on self-esteem and Big Ideas may cause us to neglect the small daily successes we need to build both. Authoritarian top-down approaches to educational reform may silence mavericks who point the way to real change. Relationships to people and ideas we already know may blind us to others who would enrich us.

If all of this sounds a bit wishy-washy, that's another problem with relational education. Almost nothing is completely true or false. Almost any idea taken to extremes can be counterproductive. Even in mathematics, there are no absolutes about how to learn.

How Drill and Kill Cripples
U.S. Math Education

16

If we teach our children merely to compute while we teach our computers to think ever more intelligently, who is going to rule the world in the future?

—Shirley Hill, President of the National Council of Teachers of Mathematics, 1980

Drill and Kill describes teaching based on mindless, protracted repetition. Mathematics is the study of patterns and the use of patterns to solve problems. Some repetition is needed to explore and reinforce patterns that the student already knows, but the oppressive monotony of Drill and Kill distracts victims from noticing patterns.

Drill and Kill is like a machine forcing children's minds to wheeze in unnatural gasps. Learning mathematics is a natural function, like breathing. Our minds need mathematics as our bodies need breathing. A good coach of singing, athletics, or yoga will nurture breathing. Similarly, a parent or teacher can foster mathematics. If mistreated, children can be so damaged that they wheeze their way through life.

Children who don't understand the underlying ideas of a Drill and Kill assignment learn to "fake it." They learn it's okay not to understand. Soon they suspect that they can't understand, so their self-esteem plummets. Judging by the math autobiographies of college students, this result of boring, unfathomable drill is very common.

Students who already understand the ideas feel clogged by rote exercises. These children learn that school is dull. It lacks the excitement of sports, TV, and gangs.

Can we change? Can you protect your own child? We begin with scientific proof that excessive drill does not help. Then we will examine some of the effects of Drill and Kill. It fosters boredom and carelessness; it fosters passive, mild, and serious rebellion; it kills joy; it disconnects math from the real world; it kills hope; and, eventually, it kills ability.

Scientific Evidence That Drill Doesn't Work

No recognized research supports the widespread misconception that mindless drill increases math prowess. Indeed, computer simulations convincingly show that excessive drill doesn't help.

Computers Don't Learn from Excessive Drill: Dr. Susan Epstein of the Department of Computer Science at Hunter College does experiments with her "artificially intelligent" game-playing program, Hoyle. She writes, ". . . it is difficult to compare how the same individual [human] learns the same skill in more than one environment; prior learning experiences are to some extent ineradicable. With a computer program, however, it is possible to learn from the beginning as often as one likes, to compare and contrast learning environments in a variety of situations, and to test the resultant skill extensively without permitting further learning. In particular, with machines instead of people, one can ask how the nature of the opposition determines what, and how quickly, a game player learns." [1]

Hoyle "learns" by playing simple games, "observing" what causes winning or losing, and basing later game-playing decisions on past observations. Dr. Epstein investigated how Hoyle learned three games: tic-tac-toe, lose tic-tac-toe, and an African game called achi. "Having learned the game" was defined as drawing ten times against a perfect competitor. (A "draw" is a game that neither player wins. Since neither player can win these three games if both players play correctly, drawing is the best any player can do against a perfect competitor.)

Dr. Epstein discovered that even after drawing ten times, Hoyle would occasionally afterward make a fatal mistake and lose a game. Her graduate student assistants suggested to Dr. Epstein that "having learned the game" should be defined as drawing 20 games or 30, instead of merely 10. They believed that more drill would teach Hoyle more thoroughly. They performed more experiments, turning off the "learning" capability of Hoyle after increasingly larger numbers of draws.

Hoyle, however, did *not* benefit from extra drill. Hoyle continued to make about the same number of mistakes after being subjected to any quantity of drill beyond the initial, helpful amount.

The computer program Hoyle does not get bored or rebellious or hopeless. Computer programs do not hate drill, but they do not benefit from extra drill. People, who are further slowed by negative emotions, cannot benefit any more.

Boredom and Carelessness

Mindless repetition is boring. Is this bad for children? You bet it is! Concentration shrivels when children are bored. Drill and Kill teaches children that math can be—indeed should be—done hastily. Haste makes waste. You can't see new patterns while you skip lightly over the Meadow.

Boredom breeds carelessness. Too bored to concentrate, students make careless errors, or simply fail to finish assignments. As they get used to not finishing assignments, it becomes a habit to let them slide. Innate motivation fades. Drill and Kill programs our children to be careless.

Decades ago, worksheets were treats. Without technology for cheap copying, fill-in-the-blanks were rare because they had to be pages in an expensive workbook. Perhaps more thought went into making them. Surely their novelty added appeal. More is not necessarily better. Copying machines proliferate worksheets like flu viruses. Some are good, but many are not. In general, they are overused.

Rebellion

Insisting that children go around and around the same groove breeds rebellion. Those who rebel may be better off than those who don't. Children who acquiesce to becoming robots are pitiable. They internalize the message that they are cogs in a wheel. They will become compliant workers some day, never reporting—or maybe even seeing—inefficiencies, dangers, or immoralities in their job setting. Do what the overseer says and you'll get by. They do their jobs.

Creative children may be especially harmed by Drill and Kill. They sense that there is something intrinsically wrong when their "specialness" is stifled. Even if they don't outwardly rebel, distrust of adults and society has been planted.

PASSIVE REBELLION

Some intelligent children respond in the easiest way for people without power—with passive resistance.

> **Passive Outer Resistance:** Stuart could have gone at his own pace in a "free" school if he would do forty problems per section promptly. He refused. He said he needed only three to learn each skill, and he had better things to do than the other thirty-seven.
>
> The teachers wouldn't let him move ahead. He just kept pace with the class, doing three problems per lesson. It was okay for him; now he has a degree from Harvard in Computer Science and a spectacular career.

Not every bright child who exercises passive resistance is as fortunate as Stuart. Too many become labeled as "incorrigible" or "slow" or "underachieving" or "lazy." The truth is that they realize that life is too precious to be wasted on Drill and Kill. If they are retained a grade because they haven't completed the assignments, they get *really* bored! Doing stupid assignments a second year makes them candidates for more serious rebellion.

MILD REBELLION

Forced practice without purpose smashes kids' egos. They cringe or they fight back.

People who see a purpose will drill dutifully, if not joyfully. Anticipated joy of music or a game makes scales or shooting baskets acceptable. In moderation, repetitive practice is okay if you see a goal.

Children know the difference between purpose*less* and purpose*ful* practice. Why should they put up with time-wasting parroting? They may complain to their parents or "act out" in school. Count your blessings if they warn you first. Let them vent.

Some schools can be truly vicious to the child who expects an education. Most are merely defensive. Even if they are apologetic, there may be problems.

> A parent relayed her sixth grader's complaint that he wasn't learning any math to the principal. "We don't have anyone on the school staff who knows as much math as Dale," the principal replied.
>
> There are "Dales" all over our country who are angry that their minds are not challenged. Their anger shows. What right does a school have to tell Dale that if he bows to Drill and Kill this year, better things may be in store in junior high? Why should he serve time, hoping that next year he may be allowed to think again?

(The good news is that Dale's mother succeeded in gaining the attention of the district mathematics director. After examining Dale, the director arranged for Dale to pursue programmed learning on a computer instead of attending math classes.)

SERIOUS REBELLION

Much sadder than Dale are the students who are bored and angry because they have no idea what is going on. Drill and Kill is killing their future—literally:

▶ Pent-up anger is dangerous.

▶ The emotional violence of Drill and Kill may lead to more tolerance of physical violence, toward others and toward oneself.

▶ Drill and Kill kills both ability and the eagerness to solve problems. ("The most endearing thing about mathematicians," said one executive director who works with us, "is that they *like* problems!") Real math education develops personal resources, giving youngsters more alternatives to drugs and violence.

▶ Drill and Kill denies students the excitement, companionship, and escape from daily problems that math can provide. Satisfactions with math may deter students from high-risk activities.

The stakes are high. Adults must stop driving angry youngsters to justifiable rebellion. We must let them develop the intelligence needed for life's surprising problems. Good math education may well be our best drug-prevention program.

Killing Joy

A friend asked me recently if I realize how often I refer to a math session by saying, "We had a wonderful time. It was great fun." My idea of an idyllic moment is sitting in my garden solving a math problem. The joy of mathematics is energizing. "Hooking" children on the pleasure of math keeps them returning to it.

> Not everyone agrees. A man once told me that when he really wants his son to remember something, he takes off his belt and whops the child with it. This surprised me. The man is a dedicated father, someone who has always been helpful and sociable in my presence. Quite apart from my antipathy for violence, I doubt that his belt helps his child remember.
>
> Drill and Kill is metaphorical whopping. "I'm going to make you remember by whopping you." That's the message of boring, meaningless drill.

I'm not saying that doing math is always pleasant. It isn't. Doing math involves plenty of failure. Failure usually isn't pleasant. Actually, it involves a tad of masochism to spend a lot of time doing math. That's no problem. Almost everyone is masochistic enough. Take a look at your own acquaintances.

Success *should* be pleasant. It isn't if the math is too easy. Like Goldilock's choice of porridge, the math needs to be just right. If we are willing and able to struggle, "just right" has a wide range for each of us. But we have to engage. Think of a baby trying to climb steps. Falls are inevitable, but success sometimes happens. What fun! It's that fun of achievement that keeps us persisting.

Killing the Connection Between Math and Daily Life

People like their learning to connect with other parts of their lives. Unfortunately, textbook word problems are too often a caricature of applied mathematics. For an example, I can do no better than to quote one of the two in a popular book: "A parking garage has 345 spaces underground and 275 spaces aboveground. How many parking spaces does the garage have in all?"[2]

Is this a question that would motivate a second grader to learn mathematics? Why would a child care? How many children know what parking garages are? Equity issues loom large in textbook word problems.

Comedians Romanovsky and Phillips spoof the way many children don't know the context for many textbook word problems on their 1994 album *Let's Flaunt It:* "Mary and Kaye are lovers. If Mary has six ex-lovers and Kaye has four ex-lovers, how many best friends do Mary and Kaye have?"[3] That apparently is funny for the "in" crowd, but most of us can only guess at the answer. Moreover, just as some of my readers may find the spoof offensive, many children find old-fashioned word problems emotionally charged and distressing. Nonmathematical issues cloud their ability to do the math. Children don't know why they should care about (metaphorical or real) parking garages or they find them so upsetting they can't concentrate on math while thinking of them.

Rarely do traditional word problems promote joy in mathematics. Well-prepared teachers, however, recognize intriguing problems in their children's lives. They can introduce problems that "hook" their particular children. Parents are in the best position to identify "relevant" math problems in their young child's environment. If they don't do it, it may not happen.

Killing Hope

Hope is essential for achievement. Some people may doubt whether joy is necessary or even possible in mathematics. But there is no doubt about hope. Drill and Kill attacks children's inborn trust that their brain can connect with whatever gives answers. It assaults the inner light of creative thinking and forces passivity. The passivity of Drill and Kill causes students to stop noticing the details that charm little children. They abandon hope of enlightenment.

Of course, not every failure is followed by success. Life gives all of us humiliating blows. We can lose our inborn faith and develop self-doubt. If authority figures subject us to Drill and Kill, teaching us we are drones, the last vestige of mathematical hope may be drilled out of us.

MAKING THE MOST VULNERABLE HOPELESS

The U.S. custom of placing children who seem "behind" in especially oppressive Drill-and-Kill situations prevents them from ever catching up. Although children who have already lost their self-confidence suffer most, both emotionally and intellectually, remedial programs are typically based on Drill and Kill. Generally, remedial students are shielded from people who know real math, often foisted off on hapless teachers who have been "downsized" out of their own specialty.

> Two teachers, Jamie's third grade classroom teacher and his Basic Skills teacher, appealed to me. Both were exasperated by his seeming inability to catch on.

"Today I did *fifteen* problems, and he didn't learn anything!'

"Slow down!" I said. "Do only *one* problem, and do it slowly enough so that Jamie understands every step."

The following day, the classroom teacher spent forty-five minutes letting various children explain to Jamie how to do one problem. The other children caught the spirit and worked hard to think up new explanations, deepening their own understanding. Jamie enjoyed the attention and tried hard. When he finally caught on, the entire class broke into applause.

The following day Jamie did a similar problem in only fifteen minutes. By the end of one week, he had almost caught up to the slower end of the class. He was no longer distinctively different. What a triumph! The teacher was also impressed with the pleasure that her other third graders obviously derived from Jamie's progress.

Helping the slowest promises acceptance for all. Humans were designed to help each other learn. We are meant to share hope.

MAKING THE MOST VULNERABLE *GROUPS* HOPELESS

Drill and Kill does the most harm not only to the least fortunate individuals, but also to the least fortunate ethnic groups. The "Back to Basics" movement, based on Drill and Kill, coincided with a continual decline in our children's mathematical ability and enthusiasm, as leading black mathematicians predicted at its beginning. Coupled with the psychic legacies of slavery, Drill and Kill did untold damage to African Americans. In 1980 one of the leading African-American mathematicians in the country, Dr. J. Arthur Jones, then an officer at the National Science Foundation, said:

> **The Effect on Blacks of "Back to Basics":** . . . I believe the "Back to Basics" movement . . . is potentially harmful to the educational advancement of Blacks in the U.S. . . . The "Back to Basics" movement emphasizes so-called "minimal competencies" for high school graduation, most of which are usually geared to an eighth or ninth grade level. Because of this focus, most of the teaching and guidance emphases will probably be limited to the achievement of only a ninth grade competency at the end of twelve years of schooling. Since Blacks populate a disproportionate number of the so-called "non-college bound" tracks, they will be a prime target . . . schools will be less likely to increase the number of advanced mathematics and science classes and will generally not make any special accommodations to increase the participation of Black students in the advanced classes. . . .[4]

A decade later Jonathan Kozol (see p. 181) confirmed the prophetic nature of Dr. Jones' words. It appears that until African-American parents insist their children be spared Drill and Kill, it will be used against them.[5]

My own work with the third grade teachers and children in Newark, New Jersey, emphasized the unnecessary waste. Although these low-income, African-American children had previously rarely scored above the 30th percentile on the standardized tests, after a few years, two classes' median percentiles rose to 60 and another's rose to 70. It didn't take a lot of my time; it could be routine if society would spend a modicum of resources for mathematically excited people to collaborate with classroom teachers.

I don't mean to imply that the standardized tests measure important math. But after these third graders were taught real math, their test scores increased dramatically. The teachers and I freed the children from Drill and Kill and showed them the magic of their own minds. They tackled new problems with hope—even on multiple choice tests!

Girls are another group socialized to follow directions. Drill and Kill, therefore, develops special self-doubt and hopelessness in them. In disproportionate numbers, they too acquire a distaste for "math." When forced to drill, many girls think, "I'm no good at this. Otherwise, I would like this, wouldn't I?" They may outwardly comply but inwardly rebel. They become bored. Since their boredom means wrong answers and they hate both the boredom and the wrong answers, they have two more reasons to think they are bad at math and give up hope.

Whatever their ethnic group or sex, children with the poorest self-image mathematically lose most from Drill and Kill and score lowest on standardized tests. Low scores generate lower self-esteem, which leads to less hope. This downward slide makes it even more likely that these kids will be subjected to Drill and Kill.

People don't try to do things they believe are impossible. Everyone needs hope. To excel at mathematics, people must hope that learning is possible. The worst aspect of Drill and Kill, however, is that eventually it may indeed make learning mathematics impossible.

Killing Ability

Joy is vital. Hope is indispensable. But there is something even more malevolent in Drill and Kill than the destruction of joy and hope. It is the destruction of ability. As the guideline document of the National Association for the Education of Young Children (NAEYC) reports, "Compelling evidence exists asserting that overemphasis on mastery of narrowly defined reading and arithmetic skills and excessive drill and practice of skills that have been mastered *threaten children's dispositions to use the skills they have acquired*" (italics mine). If someone doesn't use skills, they disappear, and nonuse (as the NAEYC document emphasizes) can result from overuse as much as underuse.[6] To people who claim that the ability is still there but unused, I would counter that *skills*, unlike joy and enthusiasm, exist only if they are used.

In Chapter 2, we noted that math ability includes joy, persistence, internal scripts, tolerance of ambiguity, and perceiving similarities among differences. We now consider how various aspects of ability are destroyed by Drill and Kill. Since we have already seen how joy is crushed by Drill and Kill, we turn to persistence.

KILLING PERSISTENCE (DETERMINATION PLUS HARD WORK)

In Chapter 2, we examined six contributors to persistence: hope, patience, productive skepticism, the ability to start over, concentration, and tolerance of frustration and failure. We already noted that Drill and Kill suppresses hope and concentration by fostering boredom, so we begin with patience.

Killing patience: Drill and Kill enforces passivity while destroying patience. There is a big difference between passivity and patience. While passivity is surrender, patience involves self-control. Patience allows the mind to relax and welcome ideas. The inner lights flicker and dance, jumping in and out of sight. If you must do rote problems that go on and on, you lose the capacity to recognize the mental bypasses that open shortcuts. Mathematical understanding could be considered a shortcut to knowledge. Enlightenment requires patience.

Adolescent loss of patience: "They give up so soon!" complain frustrated remedial math teachers. "If only they would try. I'm sure if they would pay attention long enough, they could do it. Why do they give up so soon?"

Why indeed? Suppose you had been programmed to think of mathematics as a long series of exercises, each one just slightly different from the one before, each one a boring adaptation of the previous work. If you expect each problem to be only slightly new, might you too conclude that you were incapable of grappling with harder challenges?

Killing productive skepticism and the ability to start over: In adapting to Drill and Kill, a student must turn off intellectual juices. Challenging mathematics requires the opposite attributes. Drill and Kill suffocates constructive doubt and questioning; real math requires it. Drill and Kill discourages monitoring one's own intellectual activity. Just do it! Turning back to start over again from a dead end may become a lost art. The student simply gives up.

Killing tolerance of frustration: If success keeps following frustration, a child's ability to tolerate frustration survives. Drill and Kill can destroy that.

KILLING INTERNAL SCRIPTS AND TOLERANCE OF AMBIGUITY

If each problem is similar to the prior one, you don't need to search for a new approach or pregnant questions. Metathinking is squelched.

Tolerating ambiguity is discouraged. "Do as you're told," is the message of Drill and Kill. While good math students question, doubt, and seek previously unexplored relationships, Drill and Kill encourages students to become dependent on "quick fix" answers.

Forced dependence on artificial supports: Making people dependent on artificial supports is insidious. Use it or lose it. Drill and Kill develops passivity instead of patience. Intellectual muscles atrophy, as did the abdominal muscles of nineteenth century women imprisoned in corsets. Their muscles were weakened, and they would easily faint.

> South Sea women wearing bracelets around their necks to make them longer are an extreme example of forced dependency. If the bracelets are removed, the women die because the muscles that hold up their heads have lost their power. It is the ultimate threat of husbands to disobedient wives.

Similarly, people who have had intellectual supports forced on them may become disabled when the supports are taken away. "Show me a recipe," demands the Drill-and-Kill-trained high school or college student. They may crave—even need—the security of someone else telling them what to do, one small step at a time.

Math teachers refer to this syndrome as "plug and chug."

"If you show me where to plug, I can chug."

There is more to math than that. You are supposed to figure out for yourself where to plug. You aren't doing math if you need to be plugged in. You have become a machine.

KILLING PERCEPTIONS OF SIMILARITIES AMONG DIFFERENCES

Drill and Kill, by putting similar problems together, eliminates the challenge of noticing similarities oneself, thereby damaging math ability. Students must perceive relationships themselves. Forcing them to take only tiny steps thwarts their natural desire to make Leaps. Drill and Kill forces kids to "breathe" in a rhythm that weakens their long-term ability. At first they hate it, but then they become dependent on it. After a while, they want and need their machine.

Life is not a continuum of slightly changing problems. Nor is mathematics. Both life and mathematics keep dishing out big challenges. We must retain our ability to persist at big challenges.

End-of-Chapter Exercises

An Impossible True-False Test

The following true-false questions are in the spirit of this chapter. Try arguing both sides of each statement. Whether you argue for true or false, I can argue against you.

- ▶ We learn through practice.
- ▶ Teachers know better than kids.
- ▶ People need to feel good to do math.
- ▶ Math is pleasant.
- ▶ Kids need grown-ups to learn math.
- ▶ Math can be broken down into little bits.

Life is not simple. Learning mathematics is not simple. Some statements are both true and false.

A Possible True-False Test

Here are two more with clearer answers. Do you agree with me?

- ▶ Whopping kids with a belt helps them remember.
- ▶ If a little is good, more is better. (**Hint:** Think of heat, light, hot spices, snow, cars, and consumer credit.)

Why Drill and Kill Holds U.S. Math Education Hostage

The importance of habit-formation has been grossly underestimated. . . .
. . . mastery by deductive reasoning . . . is impossible or extremely unlikely by children of the ages and experience in question. They do not deduce the method of manipulation from their knowledge. . . . They learn the method of manipulating numbers . . . by more or less blindly acquiring them as associative habits.

—*Professor Edward Lee Thorndike (1874–1949)*
of Columbia Teachers College[1]

Throughout the twentieth century, some educational gurus have justified the customs of rote teaching thoughtlessly practiced for generations. This chapter explores why Thorndike's deadening advice is still being followed.

In Thorndike's defense, we can note that he didn't foresee the delightful manipulatives and toys that now help children learn math. Thorndike did acknowledge, "The author is ready to cut computation with numbers above 10 out of the curriculum of grades 1–6 as soon as more valuable educational instruments are offered in its place."[2] If he had lived in calculator times, he might well have advocated statistics, probability, and extensive problem solving for the K–6 curriculum.

Already, by 1929 he wrote, "Systematic practice has fallen into considerable disrepute because . . . it has been overused and wrongly used. . . ."[3] He added that practice is most effective in learning "definitely arranged facts *after their significance is known*" (italics in the original).[4] Thorndike is no longer with us, and his legacy is not necessarily what he would have chosen. Would he support my tirade against overuse and wrong use of drill? Quite likely.

Some very popular current writers still seem to advocate Drill and Kill, or at least assert that practice is the "cardinal principle" of learning mathematics.[5] But learning mathematics is not easy, and to imply that practice is enough is deceptive and demeaning. It is demeaning both to mathematics and of human understanding.

If a student doesn't understand a concept, it is worse than useless to drill; it is like hammering on bent nails. If a student *does* understand a concept, it is useless to drill very much. A little practice hammers in a right idea, but beyond that it's like hammering on a flat surface.

Why does our country cling to such self-destructive customs and beliefs? Let us consider some of the reasons: misunderstanding of mathematics, misunderstanding of math education, misunderstanding memory, multiple choice tests, parental pressure, misuse of power, money, hatred of mathematics, and habit.

Misunderstanding Mathematics

Will Rogers once remarked, "You can't teach what you don't know any more than you can come back from where you ain't been."[6] Denying teachers a math education means our children won't learn math.

We previously observed that twenty-four states have no requirements for their elementary school teachers to know *any* mathematics and most of the others require only passing an easy standardized math test or taking one course. None of the states require for *graduation* the math *entrance* requirement for college in most other countries. Compounding the problem, many elementary school teachers begin college never having had any good experience with mathematics. No matter how smart and motivated they may be, their students are in trouble.

To use methods other than Drill and Kill, teachers must know enough mathematics to discern subtleties in students' questions because many profoundly philosophical issues are in the simplest childish mathematics. Deprived of any vision of true mathematics, such teachers pass anti-mathematics onto hapless youngsters. Drill and Kill provides an easy "out" if they can't respond intelligently to their pupils' probing questions.

Misunderstanding Math Education

Unfortunately, leaders in industry who work in mathematics, science and engineering don't realize how much U.S. elementary school teachers' education has been neglected. They seem to suspect that teachers *willfully* turn children against real math and science.

Corporate leaders and politicians too often think that if they mandate higher standards, the schools will achieve more. Furthermore, they seem to think that tests can enforce such mandates, whether or not teachers know math and science. Until teachers do, mandated "proficiencies" often have the opposite effect, as related on page 181.

Since Drill and Kill has become an accepted escape for teachers who can only parrot the text, colleges are relieved from the responsibility of teaching prospective teachers math. Just let them give their charges worksheets! "Practice, practice, practice." Who cares where the practice leads?

One place that it leads is to widespread cheating. Both the drillable subject matter and the types of test that measure it make cheating especially easy (see p. 229).

Don't get me wrong. I don't think that occasionally telling a "pre-addition" child that three plus four equals seven does any more harm than saying, "The cow jumped over the moon." It may do more good.

However, if children are forced to recite or write $3 + 4 = 7$ before they know about three, four, plus, and equals, they will associate numbers with being "out of it." Such children "learn" that they are incapable of learning mathematics, and that mathematics is a form of torture.

True mathematical learning is based on understanding patterns. If forgotten, the understanding is much easier to generate the second time, and vastly easier the third. Once we internalize patterns, they become part of us.

Misunderstanding Memory

I am annoyed when people belligerently assert that children should memorize their times tables, as if that were somehow central to today's educational crisis. Of course, children should memorize their times tables! And, certainly, there are an alarming number who don't.

But children who understand multiplication are able and willing to learn times tables. Given a reasonable opportunity, they will learn them. On the other hand, children who don't understand multiplication won't be able to *use* the times tables, no matter how fiercely they are drilled. A child who perceives the value of memorizing times tables can do so quickly. If you have really learned 3×4, you know you can learn 8×7. It just takes a bit more time. You could even learn 43×87 if you wanted to. But why bother? If multiplication is merely "repeated addition,"[7] who cares? Only when multiplication is seen in its full glory as repeated addition *and* rectangles *and* proportional growth *and* filling orders and various other concepts does the trouble of memorizing multiplication facts seem worthwhile.

Children who grapple with such ideas frequently will almost memorize the tables through repeated use. Some prodding from a teacher or parent may be needed to complete the task. Such prodding is appropriate *after* the child can compute the numbers meaningfully.

One third grade teacher posts a chart with the children's names and the numbers one through nine. Each child puts a star next to his or her name when he or she has learned that number's times table. Every child knows the ones, the twos, and the fives at the beginning of third grade so there are already three stars on each row. The teacher points out that although they can compute the answer to each of these multiplication facts slowly, they don't want to waste time.

When they were little it was all right to spell their name slowly. Now they want to be able to say the spelling of their own name quickly. Similarly, they want to be able to say their multiplication facts quickly.

The children work at it, and gradually the stars fill in the chart. Games and self-chosen exercises supply the fun drill. It doesn't squander the teacher's precious time. Each child knows the correct answer slowly because they understand what multiplication is. Learning the times tables with their peers also helps them learn the value—and skills—of working collaboratively.

Too many U.S. citizens seem to think that mathematics is memorizing ever-larger multiplication tables. Memorizing number facts up to ten is enough for people. Calculators and computers do large computations better than any human is ever going to do them. We need people who think.

Multiple Choice Tests

You don't have to think much for multiple choice tests that test only facts. You do have to think for more sophisticated multiple choice tests, but these are hard to make up and less common. Most standardized multiple choice tests measure only memorization of snippets.

An avalanche of oppressive, repetitive teaching techniques based on memorization has been devised by people who don't understand the beauty or power of true mathematics. They claim to help students "score high," but the learning is temporary, at best. High scores become the goal of mathematics lessons.

Easy come, easy go. Mathematical learning based on memorization soon evaporates. Human memory is like a sieve.

Parental Pressure

Teachers report that when they teach mathematics with understanding, kindness, and modern approaches, some parents object. Too many parents assume that even though they themselves didn't learn math well, their own teachers' *approach* must have been the "right" one. They want their children to get more of the same. They are uncomfortable with manipulatives, cooperative learning, writing, calculators, and exploratory conversation in math lessons.

It is, of course, impolite to criticize your child's teacher publicly for any reason. It is especially important to understand the teacher's approach before protesting. If your child is enjoying math class, it is probably harmless. It may be superb. If you feel your child is not learning at the rate you wish, continue math conversations at home.

One father was upset that his daughter did not know her times table "cold" by fourth grade. He offered her ten dollars to do so. She went to her room and emerged an hour and a half later with the job complete. He told this story angrily at a conference. Author Susan Ohanian[8] suggested several morals to this tale including:

- If they aren't too pressured, kids can learn multiplication tables in about an hour and a half when they are ready.
- Teachers need to do a better job of communicating the philosophy of academic programs to parents.

Listen to your child's teacher. Are the children learning about patterns—and about learning about learning about patterns? Home games and even bribes are not forbidden, but it is essential to keep your eye on the ball.

Misuse of Power

Some devotees of Drill and Kill seek to impose their will on children, not for the child's sake, but for their own power.

Unlike Drill and Kill, legitimate teaching of mathematics involves plenty of shared intellectual struggle, but then the genie jumps out of the bottle. The "up" moments are intermittent, but they are great. The glory of these moments bonds the learner to the teacher.

In theory, discipline arises from common purpose, but in practice every parent and teacher has some discipline challenges. Appropriate intellectual activities don't guarantee smooth sailing on the interpersonal front. They just help.

Drill and Kill, however, almost guarantees power struggles. Children will rebel, some seriously. The situation is worse if the teacher is insecure enough to *need* the rigidity of Drill and Kill. Teachers lacking confidence in their own ability in either math or discipline may fear allowing children autonomy. They can't afford children the freedom to grow. Our math classrooms should not be a battleground. Well-supported, mathematically educated teachers can have teacher-appropriate control of math classes while their pupils retain child-appropriate control of their own minds.

Teachers aren't the only adults whose need for control may interfere with children's growth. Some power-hungry school administrators must know exactly what will happen in their realm. Nothing unplanned is allowed! Fill out those plan books! Don't deviate. Secure administrators happily encourage children and teachers to explore uncharted ideas.

We each have a right to control our own minds. I applaud every child who tries to do so. I cheer every parent, teacher, and administrator who thinks innovatively but also respects children's efforts to have their own creative thoughts.

Money

Facts come cheaper than understanding. Repetitive exercises and test questions written by anonymous assembly-line workers are cheap to generate. Such exercises plod along like farm animals in the heat of summer, oblivious of the harvest to come. Similarly, test makers who need to make a profit to stay in business skimp on the expense of making and grading tests that measure deep thought.[9]

Fortunately, there are math problems and lessons written by people who feel a "vocation" to devise interesting settings for learning. Such educators take time to generate provocative enticements. Their wonderful material is available through the catalogs in Appendix C.

Elsewhere, you can identify these creative educators because creative writing is credited to the author. Repetitious drudgery is anonymous, written for money. No personal recognition is expected.

Hatred of Mathematics

Any mathematician can tell you that he or she often is told, "Oh, I always hated math!" Sometimes this is said almost proudly. Hatred toward mathematics is a socially acceptable emotion in our culture, even around mathematicians. Perhaps it explains the general acceptance of mathematical ignorance. Mathematicians frequently hear, "I never did learn math," often said with (at best rueful) pride. Who has heard someone say cheerily, "I never did learn to read" at social gatherings?

Someone who loves the subject can't help but wonder if those who denigrate mathematics by claiming it can be learned through "practice" are math haters. Some practice is needed in all subjects, but why pick on math? There aren't many of us left to defend our subject.

In math, healthy practice must *follow* understanding, to reinforce and explore it. Not so in other subjects. In other subjects, a child must *first* know some facts before gaining understanding, so practice seems more justifiable in other subjects.

- ▸ How can one learn the concepts of history without first knowing some events and dates?
- ▸ How can one classify birds without first seeing and identifying some birds?
- ▸ How can anyone learn to write without practice?

Mathematics is different from vocabulary, history, science, and writing. The "facts" intellectually follow from the concepts. A child who understands the concepts of six, seven, times, and equals can figure out that six times seven equals forty-two. Private study or peer-group games is enough.

In contrast, you can't figure out for yourself the French word for "tree," or when Columbus came to America, or how many elements there are in the periodic table. You need an outside resource. To remember the answers, you may have to drill. In *other* subjects, drill seems more defensible.

Habit

"We've always done it that way." "That's the way I learned it." "It's the policy." You've heard these arguments before.

The United States was one of the first countries to institute universal free education. Our schools grew up, often in isolated areas, before contemporary understanding of how children learn. Old habits linger. Decentralization of education has prevented national leaders from dictating modernization.

Inertia is probably the biggest catalyst of Drill and Kill. Change is difficult.

Chapter 18

What Every Parent Should Know About Testing and Grading

God give us the courage to measure the measurable,
The patience to refrain from measuring the immeasurable,
And the wisdom to know the difference.

—Fred Chichester (my husband)

. . . while our society requires product warning labels on things so relatively unimportant as personal deodorants and food coloring, no warning labels are federally required on test instruments that may determine whether someone gains employment or is classified as mentally retarded.

. . . the more we can de-emphasize the particular medium of standardized testing, the more we will be able to perceive the valid messages that this medium can convey to us.

—Walter M. Haney, George F. Madaus, and Robert Lyons
National Commission on Testing and Public Policy[1]

How can we evaluate education? Why do we try? What about grades? Are there other purposes for tests?

Your child will probably face test mania soon after she or he starts school. With luck you will resist the temptation to test your own offspring. However, everyone else talks about tests. The media worship them. How can a book on math education avoid mentioning tests? I tried, but it is impossible.

Tests drive public discussion about math, money, and much more. In a society with limited resources, their existence may be justifiable. Unfortunately, their tentacles extend beyond reasonable bounds. Recent publications by testing experts point out that recently tests have taken on a second purpose that has nothing to do with the evaluation of anyone or anything: to bring about school reform.[2] To the extent that this is true, I personally believe our money would be better spent teaching teachers more subject matter and helping them hone their teaching skills. I'm partial to math, but would be happy for teachers to learn more about writing, reading, science, social studies, singing, and Tai Chi, all of which seem more valuable to me than enormous expenditures on testing. However, I concede that tests, in moderation, do have legitimate uses.

Defensible testing is used to answer the following questions: How is my child doing? How can we better serve the special needs of this child? How can we teach better? How well is this teacher or school doing? Are the taxpayers getting their money's worth? Who gets the goodies?

This chapter begins with an examination of testing techniques and then discusses how well various tests answer the preceding questions. It concludes with a discussion of grades. Chapter 19 catalogs the flaws and misuses of tests.

How Do We Test?

Testing humans is tricky business. It really is.

Tests and grades are not objective. Human beings are judging other human beings. Biases affect human judgments, no matter how much we try to eliminate them.

Be wary of any test or test result that denigrates your child.

ORAL TESTS

In many traditions, oral tests were the norm. Early final academic examinations were public; professors asked the candidate questions while the citizens watched. Adulation or shame came quickly.

Making academic judgments in a few fatal public moments is not ancient history. Even in the 1950s, "Bolivian secondary school examinations usually were oral, a few questions from the visiting proctor to each student in turn while the others watched and prayed for mercy."[3]

Public spectacles have been long out of favor in this country as a valid method of testing. Their subjectivity is obvious. Only the listeners know what was said. Videotapes help, but are time-consuming. And how do you grade a videotape when it reveals something that can't be written on paper? Subjectively.

WRITTEN TESTS

Written tests seem more objective, especially if there are several graders. What you see is what you get; the assumption is that what a person knows is accurately reflected in what she or he wrote. We all know that isn't quite true. None of us believe that human thoughts can be accurately transmitted to paper. Trivial things, maybe, like whether a child knows what 2 + 3 equals. Even then, it is possible to draw a blank, fill in the wrong blank, or be so emotionally devastated that the test doesn't reflect what you know.

PORTFOLIOS

Vermont has been experimenting with "portfolio assessment." Students accumulate samples of their mathematics work over a long period of time and submit them for statewide evaluation. Charles Sykes claims that teachers doing the evaluating agreed on scores less than 60% of the time, and uses these data to argue against portfolio testing.[4] He intimates that overall judgments can't be fair if individual judgments vary so much.

Sykes' point is well argued, but I argue below that other measures of student achievement are no more fair. Although the reliability of the scoring may be questioned, the Vermont portfolio testing program, which involves classroom teachers in evaluating other teachers' pupils' portfolios, *can* claim considerable success in achieving its second purpose; it does appear to be changing teachers' classroom practices.[5] (Again, I must wonder aloud if the quality of teaching would have improved more happily and cheaply if the teachers were simply given more education.)

Researchers who studied and analyzed the Vermont testing program observe that society must compromise between two purposes of assessments: improving instruction and rating individual students. The Vermont portfolio system has improved classroom teachers' practices, but "the hoped-for balance between the two goals is proving elusive."[6]

"Preliminary observations of classroom instruction in Kentucky and Vermont, two states with portfolio assesessment, indicated that teachers spend more time training students to think critically and solve complex problems than they did previously," claims another review. This article adds that without investing in the professional development of teachers, no assessment program can work.[7]

LONGER TEST QUESTIONS

Deep understanding can be measured (if at all) only by complex, time-consuming problems. If portfolios collected over months are too controversial or expensive, projects lasting several days or weeks might be used, or questions that take a long time at one sitting.

Requiring students to use their own words and show their work gives a better reflection of how well students could use their knowledge in the workplace than having them merely fill in the proper "bubble"—a skill that rarely is useful on a paying job.

Many countries expect students to take over ten minutes for each problem on their national high school graduation test.[8] Questions that involve so much time cannot be fairly graded merely on the basis of whether the answers are right or wrong. Grading such questions is more expensive than grading shorter problems, where simply denying credit for every wrong answer seems more defensible.

In the Mathematical Olympiad, the international exam for the top high school students in each country, students have two days to wrestle with only six problems.[9] Rarely does someone get a perfect score.

Tests with complex questions (used in most countries) are far more expensive to grade, and tax anxiety is rampant in the U.S. As long as testing is privatized, testers must seek the least expensive way, not the intellectually most viable. Until U.S. taxpayers subsidize more expensive tests, multiple choice tests will dominate our standardized test scene.

MULTIPLE CHOICE TESTS

The United States is the only country that uses multiple choice tests. Multiple choice tests are cheap. They can be graded by machines. They also tend to be superficial. Strenuous efforts are afoot to make them more subtle, but it is not easy.

Even the best multiple choice tests can entrap good students. One careless error, which everyone makes occasionally, can be disastrous. If a #2 pencil misses one "bubble," all the answers may be wrong after that. If your eye or hand misses, you get a lower grade no matter how much you know.

Multiple choice tests are especially subject to linguistic and cultural interpretation. Thus, they are biased against any group that does not contribute to their composition. In our multi-ethnic country, some groups must be omitted from designing any one test, so test takers from these groups will be disadvantaged. Multiple choice tests discriminate against individuals who hesitate to make quick, risky decisions. They are biased against females in the sense that females with the same grades elsewhere tend to get lower scores on multiple choice tests.

Multiple choice, high-stake exams encourage memorization, which is very different from understanding, as previous chapters have indicated. Even their supporters admit that "more weight is being placed on [standardized multiple choice tests] than any human contrivance can bear.[10]

Keeping tests cheap is especially important when we give *many* standardized tests. If we gave fewer, we could have more complex questions, allow more professional time to establish validity, and use (expensive) human graders. Perhaps fewer tests would mean school districts would be willing to pay more per test. Perhaps not.

TIMED TESTS

Timed tests, whatever their form, also present equity problems. Kids from time-aware families such as mine are at an advantage. Kids with a perceptual problem are at a disadvantage.

Furthermore, timed tests fail to test the ability to sit and wait for a new, creative idea to happen. A two-day test with only six problems provides some opportunity, but to excel on tests of less than two hours with many problems, you mainly regurgitate what you've been fed. You recognize a familiar pattern and apply it.

UNTIMED TESTS

Some children with certified disabilities are allowed to work on standardized tests as long as they like. This seems unfair to those who are timed. How do we decide which children should be cut off and which may continue?

Sometimes college students are given tests over a weekend or over a week. This too has obvious drawbacks.[11]

Why Do We Test? For Whom?

There are at least four different groups who want to know test results:

1. Parents and others who have a long-term emotional investment in an individual child,
2. Teachers and administrators who are responsible for providing an education for groups of children,
3. Taxpayers, whose retirement security depends on well-educated workers in the future, and
4. Students, whose future depends upon the results of tests.

These groups do not have the same needs. Taxpayers don't need to know about individual children. Parents need to know more about their own children than grades can measure.

Let's examine six justifications for testing: informing parents, placing children in special programs, improving teaching, evaluating schools and programs, informing taxpayers, and deciding upon just rewards.

INFORMING PARENTS

Parents of elementary school children have a right to know more than tests and grades can reveal. They deserve written evaluations from their children's teachers at least four times a year and conversations whenever they like. Teachers should be accessible by telephone or in person.

Increasingly, young children are asked to play games with their families as math homework. Playing games that your child's teacher has assigned gives you a feel for both the subject matter and your child's progress, yielding genuine insight into your child's math understanding. Old-fashioned test papers coming home provide a different kind of insight. Some teachers still give worksheets, and some of these are fine. Others tell children to make up their own problems and write their solutions, stimulating both creative mathematics and thoughtful writing. If your child brings home such papers, count your blessings and thank the teacher.

In the lower grades, children deserve a good *education*. Selective college admission is not imminent, and no research suggests that taking standardized tests early improves SAT scores. Just the opposite! They can deflect teachers from teaching challenging mathematics. Elementary school teachers and children should be freed from such tests. If yours are being dragged to Davy Jones' locker by such intellectual cement shoes, do your part to get them changed!

North Carolina has led the way by making it illegal to inflict standardized tests on children before third grade. In the late 1980s Norma Kimsey, who was then an employee of the Department of Public Instruction working in early childhood education and also president of the North Carolina Association for Education of Young Children (NCAEYC),[12] led the successful drive for this legislation. She was fortunate that her friend and state senator, Charles Hipps, was the chair of the North Carolina Senate Committee on Children and Youth and that Assemblywoman Ann Barnes was another strong advocate in the North Carolina lower house. Even so, it took three years of political activity—of attending senate meetings, of making phone calls, and of discussing it almost every Saturday at the NCAEYC Board meetings—before the proposed bill passed the senate. A year later it became state law.

Kimsey now realizes that the country's passion for judging demands that something replace standardized tests, even for little children. She supports observation-based assessment that is friendly and not stressful for the children, and she admires the work of Samuel Meisels in developing such instruments.[13] Meisels recommends using (1) a checklist for observing typical behaviors,[14] (2) work samples, and (3) anecdotal records of the student's progress, all of which precede the classroom teacher's writing a summary report of each student's work.

Kimsey and I would, however, be content with the simpler method used in my children's New Jersey schools a generation ago. Teachers handwrote a short paragraph about each child four times a year on their report cards to accompany the simple checkoffs of "satisfactory" or "needs improvement." Requiring more teacher time for assessment means less time for preparing lessons and teaching.

PLACING CHILDREN IN SPECIAL PROGRAMS

Children's development is varied and unpredictable. Some children are slow at first, and then whoosh! Others are steady plodders. Others learn in spurts.

Harmful Special Programs: When the authorities threatened to put first graders who were already reading into a remedial class, their teacher protested that they were reading just fine. The tests had determined that these children weren't "ready." The teacher was told if she didn't teach "readiness" skills, the children would be transferred to someone who would. This teacher had to abandon her successful reading program to teach a set of prereading skills someone had decided first graders should "pass"![15]

Helpful Special Programs: In 1954 the first state legislation was passed in this country mandating that school districts must provide appropriate education for every child, no matter how handicapped. My mother, Bertha Francis Clark, was state publicity chair for the New Jersey Association for Retarded Children (now ARC) for the five years preceding its passage. It was a major influence of my childhood. I firmly believe that all children deserve appropriate education, and I am well aware that some children need special approaches.

Wonderful "special education" programs now exist with kind and expert teachers for children with some types of disabilities, in particular those who are blind, deaf, deaf-blind, severely retarded, or severely emotionally disturbed. Many children and families have been helped by talented "special ed" teachers when classroom teachers were overwhelmed. Yours may be too.

Sometimes, however, diagnosing children as handicapped has been used to deny them access to the opportunities that most children have and that they and their parents believe they need and deserve. Some respected studies indicate that as high as 80% of the general population could potentially be labeled as learning disabled based on one of the many criteria used by schools.[16] Many groups have expressed concern that such broad criteria are in use and worry about their effect on both individual students and our society.

Because of this, Congress passed federal law #94-142, the "Individuals with Disabilities Education Act," which mandates inclusion of children with disabilities in regular classrooms, even those with significant disabilities, *unless* their own education cannot take place effectively there *or* their presence has an unduly negative impact on their classmates, in which case exceptions are appropriate and legal. Parents of children with severe disabilities who believe the regular classroom teacher is not prepared to serve their child can request alternative placement or more support services for the classroom teacher.

Elaborate testing procedures have been devised to decide whether "placement" will be helpful. If they really lead to appropriate educational opportunities, that's fine. On the other hand, parents can well be skeptical about testing for classification. It is easier to become labeled than unlabeled.[17]

Famous Disabled Mathematicians

If your child is deaf, your family wants to learn about Charlotte Scott, the deaf mathematics professor who was also the first Chief Examiner in Mathematics of the United States College Board in 1902.[18] If your child is blind, investigate the life of the outstanding twentieth-century research mathematician Lev Pontryagin, who went blind at the age of thirteen and eventually became vice president of the International Mathematics Union.[19] Stephen Hawking, confined to a wheelchair since the age of twenty, is a professor of mathematics at Cambridge University, England. His popular books about modern physics and mathematics, written since he lost his ability to talk at age 43, have been translated into over thirty languages.[20]

Seymour Fliegel, writing from inside the New York City public school system, observed that the state gives the city $6,000 annually for each "regular" child but $12,000 for each "special ed" student. There is high incentive to classify and never declassify youngsters. ". . . a whole system has developed around the special ed entitlement: more teachers, supervisors, social workers, psychologists, and guidance counselors—a whole interest group fighting to maintain its position. These people are not even working directly on amelioration but are spending enormous amounts of time testing for certification. Furthermore, deliberately hanging onto a problem set of children keeps the total school budget higher, and that's what central bureaucrats like. . . . "[21]

Labels limit. Although there are legitimate exceptions, most children should be in the same classes during elementary school. Mary Leonhardt believes that even reading groups called "bluebirds" and "sparrows" are harmful. Everyone knows which is the slowest reading group, and children placed there lose their dreams. Leonhardt believes that methods without labeling yield more success for all children.[22]

Worse yet are the different curricula and expectations for children in "slower" groups. Research indicates that "lower-level students become locked at an early age into a permanent state of remediation;—trapped in classes that emphasize rote learning to the exclusion of ideas and analytical skills . . . [but] skill mastery is not requisite to learning more sophisticated material, . . . the early introduction of concepts and analysis can . . . facilitate basic mastery." Indeed, "Teachers are likely to view demonstrations of academic and social competence by low-track students as examples of insubordination."[23]

Individual differences merit differential treatment, not primarily by formula, but by teachers' on-the-spot professional judgment. Teachers like Marva Collins and Debbie

---

The Marva Collins Way

"The first thing I did was toss away all the reports and cumulative records. My experience had shown me that those reports were wrong more often than they were right."[24] So writes the Chicago teacher who leads a school that transforms society's rejected children into scholars. Her riveting book recounts how many children who were diagnosed as retarded, emotionally disturbed, and/or hopelessly disruptive became fine students who enjoy reading the classics.

One of her most amusing and disturbing passages[25] gives the prior records of "Arnold," meticulous accounts of how badly he misbehaved. What is missing is compassion and help for this eight-year-old. When he raised a metal pipe over her own daughter, Marva Collins reacted strongly, but she kept him in her school, where he became a diligent and devoted student.

The Meier Method

A large majority of the children Debbie Meier[26] accepted into her Central Park East schools in East Harlem were reading below grade, but 90% were on or above grade level by sixth grade. Twenty percent have diagnosed learning disabilities, but she educates them in heterogeneous classes. Well over half qualify for free or reduced-price lunches. In the school's first decade, not a single student was suspended.[27]

Meier show that successful inclusion of most children in common classrooms is possible if the school setting is supportive.[28]

Integrated classes should be small, and the teachers need support. We could have smaller classes without increased cost if we returned many special ed teachers and testers to the classroom. Classroom teaching deserves more emphasis than testing and classification. Although the ratio of school staff to students across the country is 1:9, class sizes are often greater than twenty five, and in the places that especially need small classes, often much greater.[29] Many voices are clamoring for our schools to devote more resources to actual teaching.

Tracking: The ultimate educational pigeonhole is to "track" a child so he or she will not interact academically with youngsters of different achievement levels. Officially, tracks are determined by ability, but usually achievement, ethnic group, motivation, and political pull are major factors.

Heterogeneous classes are essential until high school because young children change rapidly, and must be able to hop from one level to another. Until our elementary schools improve, however, we may need tracks within high school math. Many good people oppose all tracking, but I want some adolescents to soar intellectually, unhampered by tutoring others *in their own classrooms*. They should have other opportunities to hone their leadership skills.

Since motivation (*ganas;* see pp. 15–20) is pivotal, youngsters and their parents should have a major voice in math placement wherever "levels" exist. Occasionally, someone becomes suddenly more capable. It happens at all ages. Rules should be waived so such people can join their new peers. It can be frightening to jump intellectually, but schools should provide support and acceptance.

Seymour Spiegel started a "School Within a School" in Newark, New Jersey, because he was distressed that Newark valedictorians were flunking out of college their first semester. He gathered students who were willing to stay in school eleven months a year and take five academic subjects in each of their four high school years. Although the median reading level of the first group was fifth grade and none were reading "on level," the majority soared and took calculus at a nearby college their senior year.

Most are spectacularly successful professionals now. Five were accepted to Princeton University, and graduated four years later. Spiegel, the teachers, and the students worked hard. That's what it takes.

Tests may be helpful in determining students' capabilities, but none predicted the success of Spiegel's students. Nothing replaces hard work, imagination, ongoing alertness, and hope. We must continually question what is causing unwanted variations among children. Rebellious parents can expect to produce rebellious children. It's okay.

IMPROVING TEACHING

Educators need evaluations to educate students better. The 1995 NCTM *Assessment Standards*[30] lists four reasons for assessment: monitoring students' progress, evaluating students' achievement, making instructional decisions, and evaluating programs. The difference between the first two is in the frequency and formality of the assessment procedure.

Monitoring is a daily, hourly, even "minutely" process. Classroom teachers need to assess their pupils' progress to make dozens of professional decisions each hour. Most decisions are made on the basis of what has happened in the past minute.

Evaluating students' achievement happens at more deliberate intervals in a more formal setting. The usual format is still an old-fashioned, teacher-made test. Every good teacher knows that sense of achievement when we are convinced we have reached them all. We also know the deflating humiliation after a test covering what we thought the students had learned. Think again. Students forget. Teacher-made tests are a reality check.

Such tests are traditionally sent home to parents. Yes, they usually have a numerical grade that may be hard on the poorer students' self-esteem and the better students' social acceptance. So be it. Everyone knows those grades can quickly change. Meanwhile, they give teachers, parents, and children a sense of connection and accomplishment.

EVALUATING SCHOOLS AND PROGRAMS AND INFORMING TAXPAYERS

Taxpayers do not need to know what individual children have learned. They do need to know their money is being well used to educate future adults to care for them in their old age. Sample testing might be adequate for this purpose. Tests have been devised so that each child in a class does some of the questions, saving much time and money. The class's grade is the compilation of all its students' grades, so classes can be compared with each other, but not very accurately. More reliably, schools and school districts can be compared, but no individual child's performance is revealed.

Another measure to evaluate schools could be tests given every four grades. Goals 2000, passed by the U.S. Congress with bipartisan support in 1993, advocates, "All children will leave grades four, eight, and twelve having demonstrated competency in challenging subject matter. . . . "[31]

How might such competency be demonstrated? Will testing companies make these momentous decisions for profit? Or could we recruit enough people with subject matter expertise to volunteer to devise and grade the tests? Would the political scrapping be worth it?

As mentioned on pages 180–81, New Jersey has experimented with a "High School Proficiency Test" (HSPT) supervised by professionals living in New Jersey as part of their professional service. The HSPT test scores have been demoralizing. They emphasize what a mess we're in, and how differently we are serving students in different districts.

Political Realities

Such information is not always politically welcome. After these glum scores had been around a while, a public hearing was held by the Education Committee of New Jersey's Assembly, the state analog of the House of Representatives. The Committee wanted to return to old-fashioned tests, one way of remedying the terrible scores. Several mathematicians explained that it is important for standardized tests to measure real math knowledge, not just skills.

During my turn, the committee's chair exclaimed to the assembled crowd, which included media, "I don't care whether our children learn mathematics." I was speechless (which is rare) so after a pause he continued. "I just want them to get into college!"

What do most taxpayers want?

The lives of their graduates are the best measures of a school system's success. Perhaps asking graduates two, five, or twenty years later to reflect upon the quality of their education would be revealing, even before their lives are complete. There may be no adequate on-site measure of either good teaching or student achievement. Commanding people to devise one doesn't make it possible. We are not going to fly faster than the speed of light, no matter how much money we squander in the attempt.

DECIDING UPON JUST REWARDS

How do we decide who will be the fortunate high school graduates who enter our most prestigious colleges? Recommendations rightly play a role, but such life-determining decisions shouldn't depend solely on who had the most eloquent mentors. High school grades and class standing are justifiably major factors.

Years ago national tests were designed to compare students from different districts. Now we have two: the ACT and the SAT. They include relatively easy questions in math and English that children from educated families and upper-class schools learn to answer in elementary school. Indeed, some programs identify "gifted" seventh graders using these tests.

Unfortunately, the validity of these tests can be undermined. Very young, I discovered I could raise a youngster's SAT score about 100 points with a few tutoring sessions. Professional test preparers should do better. These days, individual tutors, private cram services, and public school courses abound, claiming to raise youngsters' scores. Often they do.

The morality of these programs is cloudy. Scoring high on college admissions tests opens paths that may have life-determining consequences, so conscientious parents and school systems may feel obligated to provide specific test preparation for their own youngsters. Some schools offer entire courses on SAT preparation. Taxpayers would get far more education for their money if that time were spent on real mathematics. It's a sad system where short-term individual goals are contrary to long-term societal good.

Why not change the system? Other countries' college admission tests measure high school level achievement, thereby providing incentives to study worthwhile topics, not just pass an old-fashioned test. Leaders at the Educational Testing Service (ETS) are aware of this conundrum and want to change the SAT test so that it measures what society needs people to learn. However, inertia is great, professionally, politically, and financially. I wish them well.

Renorming standardized tests: In 1994, the SAT, one of our country's two major college entrance exams, decided to "recenter" its scores. The justification was to put the median back at 500, as it had been in 1941. Scores had been decreasing, at least partially because a much larger proportion of U.S. citizens were being tested. Thus, the Educational Testing Service announced it would hike scores by simply adding points. By adding 22 points to math scores and 76 to verbal scores compared to previous years, the median scores returned to 500.[32]

Some education critics were outraged. How dare the testing service hide decreasing scores by simply adding points? Charles Sykes, for example, interpreted it as complicity in dumbing down our schools.[33]

A more valid criticism concerns the compression of the top scores. An 800 after 1994 in the verbal SAT was equivalent to anything from 730 to 800 in 1993 or before; in math, an

800 after 1994 includes scores from 780 to 800 earlier.[34] This muddies the results of students trying to find places in the most competitive colleges. In 1993 only 25 of the 1.8 million students taking the test received perfect scores; in 1996, 545 "earned" two 800s.[35] Competition for the top colleges is keen, and many have far less than 500 openings in next year's class. To the extent that SAT scores are valid, they are no longer as helpful to the top colleges with their admission decisions.[36]

Healthy Skepticism: My daughter, the child who was diagnosed as "incorrigible" in fourth grade (see p. 169), thirteen years later received three 800s on her Graduate Record Examinations (GRE)—perfect scores in verbal, computations, and abstract reasoning. Afterward, she said, "Mother, I didn't fill in the right answers. I gave them the answers I thought they wanted."

Some of my friends after hearing this story suggested my daughter may be unduly cynical. However, negotiating a flawed system staffed by imperfect human beings requires considerable healthy skepticism by both children and parents. My daughter knows that our family's (perhaps overly) precise use of words is not shared by everyone—certainly, not by all those employed in the mass test-making business.

Your children will take many tests. They will make a difference in their lives, for better and/or for worse. Sometimes, they may help you understand your children better. Always, however, both parents and taxpayers need to be wary of both standardized and individualized tests. What do the scores really mean? Keep your common sense in gear.

Grading

Grades are those evaluative numbers or letters that your children bring home from their teachers. They are sometimes based on tests, but they have a life of their own. Grades are part of the communication between teacher and student.

How do we determine what a student knows? How much do we believe in grades?

Should classroom participation count? My own belief is that students should be free to ask any kind of question in class without fear of grade reprisals. Others maintain that the ills of written tests should be balanced with teacher impressions. In effect, they have reclaimed the older custom of oral tests.

- ▶ What weight should be given to homework, since, at best, the teacher can tell only if it's in the student's handwriting, not who did the original work?
- ▶ What is the role of papers and projects done outside class?
- ▶ How much does punctuality matter?
- ▶ Should attendance count?

The Author's Practice

My own practice in a traditional math course is to grade only on timed written tests taken in my presence. Like every method, this has flaws. It assumes a "Formalist" philosophy of mathematics.

The "Formalists," in the early twentieth century, claimed that mathematics was simply symbols written on paper. They claimed all math could be proved from assumptions (called "axioms") by manipulating the symbols in prescribed ways.

I'm not a Formalist in that the math I hope my students experience cannot be written. Nevertheless, I am a Formalist when grading. I believe that written in-class tests are the fairest way to grade. I have no problems with the many colleagues who disagree with me; there is no right or fair way. I keep learning ways to make my tests fairer, and I admit to both my colleagues and my students that my system is far from perfect. My students have not yet made serious objections to my grading system.

After decades of working hard to make my tests and grades fairer, I am sure of only one thing: *tests and grades are not objective.* Even mine that I try so hard to make objective.

In elementary school traditional grades are worse than useless.[37] I feel extremely blessed that my own children went through elementary school at a time and place where grades were unfashionable. A system of S (satisfactory) and N (needs improvement) is adequate. Since most children received all S's, teachers sent accompanying notes to tell parents what was happening. These notes would actually tell us something meaningful four times a year. I believe there is no justification for report cards with numbers or A-B-C-D-F before eighth grade.

In middle school, my kids thrived in a heterogeneous setting with a wide variety of skill levels. The teachers urged students to compete only with themselves. "I shouldn't worry about how the other kids are doing," the youngsters would say. "I should just try to be better than I was last week." No A's lulled them into thinking they were tops. They could see more heights to climb.

Don't all children have the right to reach as high as they can without risking social ostracism by our anti-intellectual society? Not being labeled by grades allows some children to be superachievers without the peer penalties that inevitably accompany high grades in our culture. If we want some children to become intellectuals, we must stop labeling them "different."

If children are slower than average, their self-esteem is threatened by grades. If they are faster, their social acceptance is threatened. If they are average, why should they be so labeled? Doesn't every child have the right to be special?

GRADE INFLATION AND TENURE

A friend who teaches in a respectable university was told by a student, "I came to every class, and I read the text. I deserve at least a 'B.'"

In current high schools and colleges native motivation has waned, and artificial measures are needed. Grade inflation is out of control. Although elementary school children don't need grades, for adolescents it's a different matter. Let's put "C" back as the average grade. And let's put some teeth into those grades!

A summary of an individual teacher's grades for the current year and for the teacher's career might be helpful for parents and others trying to evaluate the meaning of a particular teacher's grades. So would similar data for the school as a whole. In these days of computerized grade distribution, such data should be easy to include on reports.

Students and parents who protest low grades are too often given favors. Overriding teachers who refuse to capitulate to adolescent rebelliousness undermines the system—and isn't good for anyone.

Unfortunately, there is a movement to erode teacher security just as tightening standards is badly needed. Teachers are beginning to worry that their daily bread may depend on their pandering to the short-term whims of students and parents. To give fair grades, teachers need the freedom of expression that tenure affords. Otherwise, the kids of the superintendent and school board president may get unduly generous ratings.

Tenure laws already include the right to dismiss a teacher for incompetence or moral turpitude. The courts' reluctance or slowness in enforcing educators' charges against their colleagues does not justify dismantling a tenure system that empowers teachers to give grades other than A's. Furthermore, society needs people of moderate means who can speak their mind without fear of economic reprisals. In particular, education needs such people. If the public want realistic grading by teachers, they must support tenure.

The Underside of U.S. Testing

It is meaningless to talk of "succeeding" in playing the cello. . . . There is no line with Success written on one side and Failure on the other. Children . . . do not think in terms of success and failure [until] pleasing adults becomes important. They fail because they are afraid, bored, and confused.

—John Holt[1]

. . . standards of truth in advertising related to testing are virtually non-existent.
—Walter M. Haney, George F. Madaus, and Robert Lyons
National Commission on Testing and Public Policy[2]

Judge not that ye be not judged.

—Jesus[3]

The majority of U.S. citizens call themselves Christians, and most others believe Jesus was a great teacher. Yet, we routinely violate this commandment, self-righteously judging our children with a reckless spirit that seems unchecked by any spiritual, ethical, intellectual, or financial consideration. The zeal with which children are tested far exceeds the justifications suggested in the previous chapter. Why such sacrilege, such waste of money, time, and emotional energy?

Parents need to ask that question. Why is your child being forced to take so many tests? You may not be able to do anything to prevent it. You certainly should try not to let your child know if you are distressed by how much he or she is being judged. But you *should* think about the morality and immorality as soon as your child faces test mania.

There are many problems with testing in the United States in the late twentieth century. This chapter considers the following:

- ▶ Testing is unnecessary if we already know the schools are a mess.
- ▶ Testing is intrinsically inaccurate.
- ▶ Testing puts people into unfair pigeonholes.
- ▶ All specific testing methods receive convincing criticism.
- ▶ Testing dumbs down the curriculum.
- ▶ Testing and test preparation waste time.
- ▶ Testing wastes financial resources.
- ▶ Testing practices are not consistent from school to school.
- ▶ Testing warps people's morals.
- ▶ Testing favors some groups.
- ▶ Testing can be addictive.
- ▶ Testing can kill the desire to learn.

Testing Is Unnecessary If We Already Know the Schools Are a Mess

We test and test, but everyone knows quite well enough, thank you, that most public schools are not educating our children as well as they could. Tests can verify that your kid is not learning. So? Do they help remedy the problem?

Testing Is Intrinsically Inaccurate

You can't directly inspect a person's mind. Good thing, too. I don't want anyone exploring my unexpurgated thoughts! The best tests are an attempt to "see through the glass darkly" into other human minds. Even with good tests, however, there are many pitfalls.

John J. Cannell, M.D., discovered in the mid-1980s one of the most curious glitches in standardized testing.[4] With the help (only) of his nurse, his lab technician, and his X-ray technician, he contacted every state education department. He discovered that all fifty states report being above average. More than 90% of the country's school districts have above average performance. Over 70% of the individual children tested as "above average." Lake Woebegon flourishes.

How can this be? It's simple. If "average" is established at one time and then children take the test over and over, their scores will increase. If teachers have seen previous versions of a test, they can "train" most of their children to "score high" on that particular test.

Children become like trained seals. Jumping through tests. Any fixed test can be "beaten." We have a country full of administrators, teachers, and children beating tests. The tests invite us to deceive ourselves about what is valuable and true.

Testing Puts People into Unfair Pigeonholes

The Japanese believe it is impossible to predict whether someone will succeed in college until eleventh or twelfth grade. We subject little children to standardized tests and take the results very seriously. If one's fate is revealed in a test score, why work? Parents must consistently doubt disappointing scores; otherwise, our system will grind down most students.

Jay Mathews has written a scathing indictment of the Los Angeles tests to identify "gifted" second graders. "[Escalante] proudly displayed his calculus roster, where gifted students were always a minority. . . . by the time AP calculus classes grew large enough to provide a roughly accurate sample, gifted students proved themselves the least successful of any identifiable group. They made up only about 20 percent of the calculus team, and only about half of them passed the AP examination. . . ."[5]

Mathews shows the injustice of early testing for giftedness, first from an administrator and then from a student:

> The system made it difficult to test new students later on, and forbade retesting children already on the gifted list. "If we retest them and find that they are not achieving at that level anymore," said Bennett with a slight smile, "we don't get money for them. Besides, it costs seventy dollars just to test somebody."[6]

[A calculus student, Orozco] remembered the test they had given him in the second grade: make or break, genius or dunce. There was a picture of a man on a bicycle beside a picture of a truck. The truck was smaller, at least in the picture.

"Which one do you think is heavier?" the lady asked.

Orozco recalled being overwhelmed by the possibilities. The bicycle might have been made of lead. The truck could be paper-mâché. The man on the bicycle resembled Fats Domino; that would certainly make a difference.

"The man on the bicycle," he said. He assumed that was why they never put him in the gifted program.[7]

An academic study of "ten racially and socioeconomically mixed schools undergoing detracking reform" reported in the prestigious *Harvard Educational Review* concludes that "the real stakes . . . are not academics at all, but, rather, status and power."[8] It reports one "White, upper middle-class father" was "outraged" because "'precious' state resources for gifted and talented students were being spent on 'non-deserving' students—many of whom had higher middle school achievement test scores than the students who had been identified by the school district as gifted many years earlier."[9] It's seems odd to decide whether a child deserves special opportunities in middle school based on a test in the primary grades, but the fact that this does happen magnifies the preschool impact of parents.

(My moderate position in this debate was explained in the previous chapter. Some children need and deserve special classes, but most young children flourish best in untracked, self-contained classrooms with one well-prepared, well-supported teacher not plagued by "pull-outs." Until our elementary schools really serve all youngsters, however, high school math classes must cater to the preparation and motivation of students. For example, many youngsters are ready for calculus in twelfth, eleventh, or even tenth grades, but not all. Alternative math courses are justified.)

All Specific Testing Methods Receive Convincing Criticism

Some people clamor for "Outcome-Based Assessment." A full chapter attack on this approach is in *Dumbing Down Our Kids* by Charles Sykes.[10] He believes that Outcome-Based Assessment is "slippery" and of dubious value. Oral, written, timed, untimed, and multiple choice tests were challenged in the previous chapter.

The predecessor to Outcome-Based Assessment was "input assessment"—based on class size, books and computers available, salaries of staff, parental income and education, physical facilities, and amount spent per pupil. This too has flaws, and some rich people actually deny that lack of funding affects the obviously different effectiveness of schools available to poor children. Some people whose children and grandchildren are receiving the most expensive education available in the nation complain that money can't buy quality for others—so don't raise my taxes, please!

Testing Dumbs Down the Curriculum

If the stakes are high, teachers will "teach" to tests. Indeed, this is one purpose of testing: to promote uniformity in teaching. Alas, the floor may become the ceiling.

Unfortunately, here and now, it's worse than that. The very existence of our tests is dumbing down the curriculum. Until we learn to devise (probably expensive) tests that reflect true mathematical learning, and we *revitalize the education of teachers,* testing will continue to depress standards of mathematical achievement in this country.

Using Drill and Kill before current standardized tests, a teacher can raise scores. I fill you up and you spout it back. (Then you forget. Then you conclude you are stupid.) But people don't retain hastily memorized, meaningless facts.

Dr. Cannell wrote, "I decided to present myself to a test publisher as a superintendent of schools from a small southern Virginia school district. I called a publisher and expressed interest in purchasing the company's standardized achievement test. . . . Almost immediately, I was talking to a saleswoman who implied that our district's scores would be 'above average' if we bought one of their 'older' tests! She further intimated that our scores would go up every year, as long as we 'didn't change tests.' " [11]

Charles Sykes, one of the loudest voices screaming for educational "accountability," has written a scathing chapter indicting U.S. standardized tests, both those of private companies and those of some states.[12] Given the current privatization spree, one of these test companies probably will win the contract to develop national tests if we adopt uniform national testing. There are over 500 companies to choose from,[13] but no adequate way to judge the quality of what they provide.[14] How do Sykes and team think we might obtain "accountability" for students when there is none for test makers?

Even the best test will interrupt student learning. As Marilyn Burns has remarked, "Plants grow better if you don't keep yanking them up and looking at their roots." [15]

Testing and Test Preparation Waste Time

In most U.S. schools all education stops for about two weeks before each standardized multiple choice test. Teachers teach to the test or, in some cases, the test questions themselves. For two weeks our children are drilled for testing, as if that were the goal of life. A 1990 study[16] by Mary Lee Smith of Arizona State University found that the average U.S. elementary school teacher spends about 100 hours per year (four weeks) preparing for and giving standardized testing.[17]

Big posters on school halls urge kids to do well on the upcoming standardized tests. Conversation in the halls and teachers' room centers around them. Real education takes a back seat to memorizing quick facts that may help children raise their scores.

As recounted in Chapter 15, some school systems in urban areas pressure teachers to abandon teaching serious subject matter in high school altogether for test preparation. Others distribute books with titles such as *Scoring High* and expect elementary school teachers to spend a period a day all year long drilling children for standardized tests. It makes the city look better to the media and politicians. Too bad if the kids fail to get an adequate education.

How many adults use multiple choice test-taking skills? What justification is there for this ritualized interruption of our children's education? What does it tell them about our society's ethics (and religion)?

Testing Wastes Financial Resources

The testing industry consumes too much of our "education" dollars. If you care, you too should use interlibrary loan to read an amazing book *The Fractured Marketplace for Standardized Testing* by three testing experts.[18] (Why isn't this book in every public and university library?) Its authors emphasize the impossibility of obtaining correct data about an industry "shrouded in secrecy,"[19] but in thirty pages of painstaking explanations they estimate an upper and lower bound on the annual costs of standardized testing in the late 1980s for all U.S. public school systems from K–12.

They conclude that a minimum estimate was about $311 million, but if we include the costs of professional salaries and students' loss of education, the total rises to $22.7 billion. They observe that the number of standardized tests had been increasing by more than 10% a year for several decades. Assuming a similar acceleration into the 1990s (which seems likely) and multiplying by an inflation factor of 1.42 (from government data), a straightforward computation indicates that in the late 1990s we are paying at least $1.2 billion for the direct costs of K–12 testing and probably more like $87.6 billion for the total costs.

Haney, Madaus, and Lyons' upper estimate for the costs of testing was about 13% of the total budget for K–12 education in the late 1980s. When I update their estimates for only five years, I conclude that in the early 1990s (the most recent years for which total budget numbers are available), standardized testing cost 20% of the entire U.S. K–12 public education budget. There are two reasons for the increased percentage: (1) greater amounts are spent on testing each year and (2) taxpayers are becoming increasingly stingy with total educational financing. (Increasing the numerator while decreasing the denominator makes a fraction larger.)

Understanding Large Numbers

How can ordinary mortals understand such lofty numbers? The *lower* number is more than ten times the budget of Montclair State University, which provides fine higher education to 13,000 students. It is about the same as the much-discussed debt of the United States to the United Nations in 1997, which may determine its survival.

The upper estimate is about one third of our current military budget. If divided equally among the approximately 84,000 schools in the country, it would be a windfall of over a million dollars per school. Divided among the roughly 47 million K–12 students in the country, it is over $1,800 per child.

The enormity of the cost problem can be seen another way. Suppose we accept Dr. Smith's estimate that U.S. K–12 students spend an average of four weeks a year taking and preparing for tests. My observations indicate it is much less in good schools and much, much more in urban areas, so I believe this is a plausible *average*. This is about one ninth (or 11%) of the school year, during which time almost all local "education" costs are spent on testing. On top of that, there are the full-time salaries of both local and state level people who choose, administer, and counsel about the tests, plus the costs of test booklets, answer sheets, and grading. Thus, we reach 13% without any consideration of the effect of testing (instead of educating) on the children's future—and our old-age security.

The only thing that seems really clear is that nobody knows what we spend on testing. Alas, there seems to be no accounting for—or accountability for—the costs of testing. The K–12 budget analysis presented by the U.S. Department of Education (on the Internet at www.ed.gov) includes *nothing* for "testing." Apparently, the costs are all hidden in salaries, administration, building, and counseling costs.

"In 1959, teachers' salaries accounted for 56 percent of the operating budgets of American public schools. By 1989, the percentage had dropped to 40.4 percent." [20] With so much money spent on administration and testing, less money is available for teachers. Both class sizes and the incentives to enter the profession suffer. Furthermore, if the money spent on standardized testing of small children went to educating their teachers, our children's achievement could soar. Even a three-week summer program for teachers can bring about amazing changes in their pupils—both in the children's command of real math and in their scores on old-fashioned standardized tests. Imagine what might happen if we invested significantly in present and future teachers' mathematical knowledge!

We need more "truth in reporting" on the finances of standardized testing. How much does *your* district spend to acquire and grade standardized tests and on the salaries of those who choose and administer them? What proportion of your child's school year is spent taking and preparing for standardized tests? If you take this fraction of salaries and build-ing maintenance costs, how much of your local taxes are going to testing? How much of your state taxes? How might you find out?

Testing Practices Are Not Consistent from School to School

Once I spent a day with a group of third grade teachers in one municipality. Two came from each school, spending a full day with a college math professor. It was wonderful. Teachers really *want* to learn math.

However, when I returned from the midmorning break, one of the teachers showed me a problem from the standardized third grade test that she had given to her pupils the previous week.

"None of us can solve this problem," she said. "Can you show us how?"

This was itself interesting. *None* of those motivated teachers could solve one of the problems on the third grade test. As I was considering how to teach them how to teach that type of problem, another teacher asked the person who had handed the paper to me, "Where did you get that copy of the test?"

"Oh, everyone in our school was given a copy." The second person was indignant.

"We were allowed to have the tests only for the night before," commented a third.

"We got them only the morning of the test and had to return them that afternoon," complained a fourth.

"We had them only during the test itself," said yet another. The sense of victimization was growing in their voices.

"We weren't allowed to see our grade level tests at all. We all gave tests to kids in other grades."

There was a horrified pause.

"Can I see that paper? My school will be giving the test Friday."

This is a true story, and plenty of other tales indicate that it is typical of the entire country. Even in the normal course of events (which testing isn't), people don't do exactly as they are told. They stretch the envelope a bit.

The only way we can minimize (not avoid!) the inevitable problems of administrators and teachers giving tests to their own students is to have third-party proctors give un-opened tests at exactly the same hour, as is done for college admissions tests. This is very expensive and time-consuming, but if standardized testing isn't worth doing right, why bother? The risks are high.

Testing Warps People's Morals

Current U.S. standardized tests foster cheating. Outright cheating (by my definition) is more common than I would have guessed before I began roaming the public schools. When asked about the cheating in their schools at standardized test time, most teachers just roll their eyes. When pushed, they may say, "You don't want to know." Others look like they were just stabbed and change the subject.

One replied, "Oh, I don't think there is much cheating in this school." She paused. "Nothing worse than pointing to certain answers on the children's papers and suggesting that they do that one again."

Dr. Cannell did another study that made him suspect outright fraud by school staffs is common while administering and handling standardized tests.[21] However, Mary Lee Smith concluded that teachers regard the tests themselves as fraudulent and "see fundamental discrepancies between true educational attainment and information conveyed by test scores."[22] She concludes her report by saying, "You can't cheat a cheater, which from the teacher's point of view is the standardized test itself and those who use its result fallaciously or for inappropriate purposes."[23] Cannell and Smith observed similar facts but interpreted them differently. Whatever the interpretation, these facts challenge the widespread faith in standardized testing.

Stories in the popular media—TV, magazines, and even the most highbrow newpapers—about cheating in schools previously praised for high test scores are becoming nearly routine.[24] We can scapegoat the principals involved and feel smug and superior, or we can address the true culprits: counterproductive teaching methods and testing philosophies. No American can afford to feel smug or superior about these.

Shouldn't we reconsider the system? If almost all teachers have so little faith in it (as Dr. Smith and I observe), perhaps there is some validity to their views. Who is the watchdog for the validity of standardized testing? Some who have studied this question believe there are none—and none in sight.[25] They believe our children are too often the victims of incompetent test makers—or worse.

On the other hand, not all evidence of cheating indicates the real thing. Students taught in the same environment will tend to make the same cluster of mistakes; this is especially true if they have had little comparable stimulus outside the school environment. A fascinating exploration of one of the most widely publicized accusations of cheating occupies thirty-six pages of Jay Mathews' book *Escalante*.[26] It was resolved by students' taking the test again while being watched by special proctors from Educational Testing Service—and passing. A fictionalized, much less thought-provoking account appears in the popular film *Stand and Deliver*.

Testing Favors Some Groups

Women statistically get lower scores on multiple choice math tests than men who get the same grades in math. Study after study reflects this stark fact.[27] MIT led the country in facing it realistically by changing its admission policies. Now women's grade distributions are similar to men's, and the overall achievement is greater.[28] Previously, the women MIT admitted earned significantly higher grade point averages than the men.

Nobody knows why multiple choice tests are biased against women, but my guess is that it results from girls being socialized to be careful and accurate. Studies show that in our schools girls are punished for careless errors and risk taking more than boys. Multiple choice tests require quick guesses.[29]

Dr. Peter May, former mathematics department chair at the University of Chicago, reports that the Graduate Record Examination (GRE) scores have practically no predictive value for either sex in determining who will be a good Ph.D. student. How *could* a multiple choice test predict who will be able to prove theorems that nobody ever knew before? It's an entirely different activity. The Chicago mathematics department looks for other predictors in choosing their graduate students.

The differences among ethnic groups in test scores is even more startling. I was impressed by a black student in one of my graduate courses and suggested that she apply for a Ph.D. program. Because she got less than 600 on her math GRE, some of my colleagues urged me to stop supporting her, arguing that she wouldn't succeed in a competitive doctoral program, and I was only hurting her by saying she could. They were wrong. She went to a university where the typical scores were over 750 and did well.[30]

There is much evidence that members of some ethnic groups are vastly better at math than their scores on multiple choice tests indicate. It is not a sign of "lowering standards" to note this when choosing people for limited goodies. However, it is fraught with ethical problems for those of us who are grading Formalists. Personally, I try very hard to help my minority students overcome test biases against them. Sometimes I succeed, but sometimes I don't.

> Koko, the gorilla, who knows 1,500 words of American Sign Language, soon gets bored when psychologists try to give her IQ tests. Her human companion claims this is why Koko doesn't score higher. Why should she try? She prefers playing when serious psychologists ask questions. They conclude she isn't intelligent, but her daily companions believe she knows far more than she shares with her pompous interrogators.

The results can be tragic for humans who refuse to play the game. Children who are accustomed to being asked questions by people who already know the answers are better prepared to score high on tests. Playing "quiz" with preschoolers prepares them to play adults' silly "games." Be sure not to fret when your child's answers are wrong. *You* are not testing your children. You are only teaching them to play a game—and giving them some happy attention. With luck, they will play the game happily when it's serious. If not, you've tried.

Testing Can Be Addictive

Strange but true. And to the extent we become dependent on tests, we aren't prepared for the real challenges of learning and of life. Forced dependence on artificial supports is insidious.

> I was good at taking tests. What a sense of betrayal that first May after the birth of my first child! I had no tests. I had worked very hard that spring, but there was no proof of accomplishment. Although I was laughing at my own emotions, it seemed strange, almost unfair, that there would be no final exams and a summer off. Regular testing provides poor programming for full-time parenthood.

It can start much younger. Some second grade teachers report that some of their pupils already say, "Stop talking about the math. Just fill in the paper. That's what you're supposed to do!" How sad. Already, they have lost the joy. They believe filling in the blanks is the purpose. Of what?

Testing Can Kill the Desire to Learn

When tests kill the desire to learn, they have killed the goose that lays golden eggs.

Failing a test is not fun. If students associate learning with tests and failure, they may lose their inborn enthusiasm for learning. Most tests are set up to determine what students don't know—in other words, to assign failure. In *Parents Who Love Reading; Kids Who Don't,* Mary Leonhardt argues that tests play a major role in destroying the love of reading that many children in her (upper-class) town have when they enter school.

Good grades can be a turn-off too. Being different is not comfortable, no matter how you are different.

> **Pretending to Be Dumb:** A teacher told me of a boy whose parents decreed he would repeat sixth grade unless he raised his standardized scores radically. In one year his scores jumped from the 30th percentile to the 99th. He had been sabotaging the tests because (1) he feared his peers' reaction if he were certified smart and (2) he didn't want to be put into a gifted class. Only extreme parental pressure kept this child from intentionally pretending to be much less able than he was.

Tracking children is one of the villains of this story. Why should a child have to worry about placement? Shouldn't sixth graders be free from that anxiety?

The damaging effects of standardized tests on children's minds and souls are widely recognized by teachers, but the tests continue without convincing proof that they are beneficial. Human innovations "take on a life of their own, and restraints on how they are used over time can be very tenuous indeed." [31]

The urge to test is partially a result of a misunderstanding of what math can do. People's achievement and abilities can't be ordered in a line. Labeling individuals with numerical test scores is misleading, at best. It can be very damaging.

If we *must* judge other people, we had better know why we are doing it and what good might come from it. And we had better hope (and pray, if we pray) that more good than harm will result from testing.

Meanwhile, we must do our best to protect our own children and those around us from anti-intellectualism by finding them friends, classmates, and teachers who want to learn. Some of us never outgrow our need to learn. It is a great blessing.

PART V

TWEAKING THE MACHINE

Together We Can Do Much More: Creating a Climate

20

Individually, we can do so little. Together, we can do so much!

—*Helen Keller*[1]

Parenting is too big a job for one or two people. Today TV and computers are major features in most homes and greatly broaden families' perspectives. Don't overlook the TV programs that feature math.

But TV and computers, no matter how excellent, can't replace warm, live people. Even if your family can afford a wide array of computer games and your child communicates with people all over the world on the Internet, don't confuse that experience with interacting with nearby friends and loved ones. "The Net" can relieve loneliness and enrich perspective, but we still need people physically nearby to share our activities.

Although the traditional communities of aunts, uncles, grandparents, older siblings, and cousins had dispersed by my childhood, neighborhoods provided stable sets of extra parents as numerous families raised their children together. Betty's mother led us in handicrafts. Connie's mother was the musician. My mother sponsored my Girl Scout troop for the astronomy badge—because I wanted it, and in those days, it would have been "undemocratic" for one troop member to get a badge alone!

Today's unstable families and mobile neighborhoods make it ever more important for groups of parents to band together into "extended families," even if they must be temporary. The good news is that parents who try to organize supportive groups are often welcomed by their peers. Furthermore, there are many youth organizations already in existence that welcome parents who suggest a specific activity. Whether you start a group or join an established organization, finding a parent group that values math is extremely helpful.

A Group for You and Your Preschool Child

If you are living in a typical U.S. community, you may want a congenial playgroup for your child and discussion group for parents even when your child is a preschooler. Anti-intellectualism is so accepted in our society that well-meaning neighbors may question whether you have a right to stimulate the cognitive development of your child. Even if you are consciously convinced you have that right, the dour comments of people around you who associate learning with misery may cause more trouble than you realize.

Some folks may even try to convince you that you don't have the *ability* to teach your child. Can you count? Recognize numerals? Can you add? Then you can stimulate your preschooler. Just remember to keep it fun. Nothing specific is urgent.

Whatever your current mathematical status, both you and your child can benefit from a group that shares your goal of high achievement with joy. We all harbor a desire not to be seen as a fool. We don't like to be accused of being misguided. If we are teaching our child math, it is unsettling when people suggest we are foolish or misguided.

A support group is invaluable. If your child is involved in a nursery school or day care center, that is the obvious place to first seek parental companionship. If not, perhaps a nearby nursery school, day care center, or parents' center can direct you to like-minded parents. If they try to discourage you, perhaps the public schools can help you find a like-minded group of parents. Many support groups have been started by one letter to the newspaper.

If you have access to the "Net," you can start there. A Net group will not provide eye contact for you or the companionship for your child that a local support group will, but it may be a way to reach people near enough to meet. It may give you lots of ideas and inspiration.

If you can find at least one or two congenial families nearby, hold a meeting in your home. Major changes start with a few people. What would you do at such a meeting? Start by going around the room saying names, describing your families, and telling why you came. Then go around the group again to trade ideas and tell stories about your adventures with your own children. Throw out questions and consider them together. You know that the others are not "experts," but they may see options that you missed. You will learn plenty from each other. I promise!

Remind each other that math should be fun. In our Drill-and-Kill-possessed society, this needs reinforcement. It is the all-important message. If you lapse into Drill and Kill, you risk undermining everything else. Remind each other that anything specific you accomplish before your child begins school is "gravy." However, you will want to share types of fun drill that your children have enjoyed.

If the conversation rambles to the atmosphere in your child's designated elementary school, that's fine. Now is the time to investigate its quality and, if need be, to use your collective influence to affect your children's formal education.

Your School's Culture

When U.S. children are asked what they want most, they mention money, things, and the ability to go to another planet. Asian children express a wish to do better in school and to be able to go to college. Teachers and parents who want high-achieving youngsters must buck our culture.

The most obvious place to begin is the PTA or PTO. As a newcomer, you will want to first observe the tone, the organization, and the leadership, but as you become involved, you will want to help the group support inspiring education. The school's mathematics program should be no exception. If your school system has not yet begun to implement the NCTM *Standards,* now is the time.

Toward this end, 26,000 kits called *Math Matters* were sent to local PTAs in 1989. The kits include seventy at-home activities for parents who want to help children with shapes, sorting, number sense, measurement, estimation, and graphs. They also had a videotape designed for PTA meetings and back-to-school nights and two posters touting the value of math skills. Perhaps you can find the closet that contains your school's kit and air it for current parents. (Alas, it is no longer available elsewhere except for pieces on the PTA's website.) Former national PTA president Ann Kahn publicized the kits saying, "Probably the most important thing parents can do to help their children with math is to change their own attitudes. . . . How many parents discourage their children from taking advanced math by telling them that math is too hard or that boys are naturally better at math than girls are?"[2]

Whatever the current stance of your local parent-teacher organization, ask to borrow the school copy of the NCTM *Standards* and *Everybody Counts* (see Chapter 11 and p. 266). If the school isn't lending out copies, that should be remedied pronto. If the principal claims poverty, what more urgent matter for PTA fund-raising? Perhaps you can donate copies yourself. A subscription to the NCTM's *Teaching Children Mathematics* should grace the teachers' room table. If it doesn't, perhaps you can help remedy that, too. (As mentioned on pp. 166–67, anonymous memberships in NCTM might be provocative gifts.)

Investigate what the school is doing to integrate modern topics and teaching methods into the curriculum. Is the principal trying to bring the school more in line with national standards and policies? If so, you and other enlightened parents will want to do everything you can to help.

For significant math achievement in your child's school, the principal must care. If you go for a conversation, you might want to take this book along. Ask about making copies of all three NCTM *Standards* documents easily available for parents and teachers.

One activity for your PTA or PTO might be a meeting featuring a mathematics consultant who shares modern approaches to mathematics. During the discussion period, the consultant, parents, and teachers would discuss the next steps in your school. Nineteenth-century mathematics is not good enough for twenty-first century adults.

Sometimes it is crucial to have a group of parents poised to speak out for educational values in their school. The resistance to improvement can be mind-boggling. When a Houston suburb was considering joining Ted Sizer's Coalition, "angry parents forced them to back down when it was learned that *the school football team would be unable to practice during class time*" (italics in original).[5]

Other times parents eagerly latch onto new ideas. One success is a teacher's giving a new math game each week to the parents of children who seemed "behind the pack" and encouraging parents to play daily and adapt the games to the needs of the child. The results were gratifying for the children, and some of the parents even admitted that they too finally understood the primary grade math curriculum after playing the games with their child. Some even played them alone[6] when their child was in school to keep up!

State math conferences can provide an exciting rush of ideas and inspiration. Perhaps you can invite your child's teacher to go with you. The displays and presentations there may give you both a major "lift."

Math Activities for Schools and Other Groups

Math fairs are an effective way to spread the joy-in-mathematics word. One classroom can make up math displays and invite either the whole school or others at the same grade level. Some schools devote one day a year to a "math fair," where each classroom prepares its exhibits and everybody travels to see the others. Sometimes these "Mathematics Olympics" include parents, teachers, and students creating many math activities and enjoying them together.[7] A competitive aspect for each grade may add excitement, but some schools decide the activities should focus on intrinsic pleasure. Some districts have a centrally located math fair; Newark New Jersey's math leaped forward when it instituted city-wide annual math fairs.

Children enjoy dramatizing math problems, making up their own skits and stopping in mid-skit for the audience to supply the answer to a particular problem. I've seen sixth graders present such a skit about selling Girl Scout cookies, and another about how long tropical rain forests will last at present "harvesting" rates. A great time was had by all. Neighborhood children could make up skits for each other and for their families.[8]

Another valuable activity is sending out youngsters to interview members of the community about how they use mathematics in their work. The children can then design and make a poster about their findings for display at the town library or other public place. If your local newspaper ran a feature about the study and its accompanying display, it would help the community understand why taking math seriously is important to everyone. Some algebra teachers have led their classes in such community-wide research projects about the uses of algebra, with resounding success. Younger children could focus on arithmetic.

"Math in the Park" is a program in which high school students make up group games that teach math and share them with younger children. This concept is still in its infancy, but early trials are promising.[9]

All of the above are possibilities for a PTA program. Parents can enjoy skits, applications of math, possible careers, ways of teaching younger children, and other demonstrations of how math has become real to their children. It can boost attendance at PTA meetings while helping to create a math-sympathetic atmosphere in your child's school.

If you have a math-based career yourself, your own child's class will enjoy hearing about it, and it will be fun to create a lesson that gives them a glimpse of your daily life. Once you have done so, other classes may appreciate hearing from you too. Whether or not you can do this, look around to see whether other parents could share the math in their career with your child's class and other classes in the school.[10]

Family Math

If your school doesn't yet have a "Family Math" program, now is the time for your principal and PTA (or PTO) to organize Family Math sessions for a group of families, including your own.

"Family Math" is a national program that runs a series of programs on evenings or weekends for families to enjoy math activities together.[11] Every child must bring a "parent," and every adult must bring a child, although a one-to-one correspondence is not necessary. An only child may bring both parents, and a single parent may bring all the children in the home. A child can bring an aunt, uncle, grandparent, or special adult friend. School-based Family Math programs generally involve weekly sessions.

Children and adults share cooperative math activities. Many ideas about how to continue enjoying math together at home are suggested. The book *Family Math* began the program and provides seemingly endless variations on math games that families can enjoy together. It is available in many of the catalogs listed in Appendix C if not locally.

Family Math sessions are usually led collaboratively by two local elementary school teachers. Enlightened elementary schools send pairs of teachers to national centers for a three-day training session, and invariably the teachers return full of exciting ideas for stimulating families to enjoy math together. If your school is not yet thus enlightened, ask your principal to send two eager teachers; most states now have training programs.

Family Math is a social opportunity you don't want to miss. It's great fun, and a perfect opportunity to meet others with your values. No matter how mathematically inept you feel, you will enjoy Family Math sessions. No matter how mathematically astute you are, you want to take advantage of the intergenerational pleasantries. Whatever your feelings and math background, the people you meet in Family Math classes will be your allies in developing a math-supportive community for your child. Over the years this may be helpful in integrating the "work" and "play" masks of mathematics. If your child's teacher resists teaching real math, this group may have fertile ideas about how to reach that particular person.[12]

Other National Resources

You don't have to reinvent the wheel. If you or someone in your group has access to the Internet, ideas are pouring forth there faster than one can read them. The websites in Appendix D are good starters, and they will refer you to others.

Whether or not you can surf the Net, you *can* tap into a variety of national organizations that want to reach parents. Appendix B lists organizations offering many different types of help for parents who are trying to raise math-happy children. Many people do care!

Traditional Group Math Games

Any youth organization can include math games or more serious math in its program. The geometric patterns of square dancing and other types of folk dancing are inspiring and fun. Two games that stimulated my math interests when I was a child were Buzz and The Thread Follows the Needle.

Buzz involves counting around in a circle, each participant saying a number until someone makes a mistake, at which time they drop out of the circle. The last person to survive "wins." The interest is created by saying "Buzz" instead of any number containing a seven or divisible by seven. Thus the counting goes 1, 2, 3, 4, 5, 6, Buzz, 8, 9, 10, 11, 12, 13, Buzz, 15, 16, Buzz, 18, 19, 20, Buzz, 22, 23, 24, 25, 26, Buzz, Buzz. . . .

The leader could insist upon a maximum time of one second or two seconds, after which each player is eliminated, but usually the rhythm of the game catches all the players into a snappy pace, regardless of whether they are prepared for their "number." (If there are only a few people in the circle, it's fun to reverse the direction of the rotation with each "buzz." With a large group, though, reversing means most people will never participate.)

The Thread Follows the Needle is an active game played while the group sings a pleasant song whose words are "The thread follows the needle, The thread follows the needle, In and out the needle goes, While Mother mends the children's clothes."

The children begin by standing in a row holding hands, all facing the same direction. Two children on one end of the line raise their clasped hands to become a "bridge." The child on the other end of the line, therefore, becomes the "needle." The needle walks forward in a semicircle, leading the thread of children, all of whom still hold hands, through the "bridge" of two hands held high, and out to their original position. The second to last child is now standing with arms crossed, facing the opposite direction from all the others.

That child and the next adjoining one (away from the previous bridge) raise their hands to form another "bridge." The needle then leads the thread through the next "bridge," and the next, each time leaving another child standing with arms crossed. The geometry of The Thread Follows the Needle intrigued me in the primary grades and, in retrospect, sowed some of the ideas and interest for my Ph.D. thesis.

Established Groups Outside School

The programs of Boy and Girl Scouts, Boys' and Girls' Clubs, 4-H clubs, "Y's," and religious and other organizations can include math games with much pleasure and no pain. Community groups can sponsor pilot projects until the public schools realize that parents and children will respond to challenging math programs. Parents and sympathetic citizens can use their success to hasten school leaders' awareness that serious mathematics should be available to all children.

Girls' Clubs in collaboration with the American Association of University Women are offering activities to remedy the sex imbalance in mathematics. I am glad for these activities, but the math achievement of U.S. boys is nothing to brag about either.

Tutoring services are sometimes sponsored by churches. When I was a graduate student in Philadelphia in the '60s, a church near the university ran after-school tutorial programs for neighborhood children that involved twice as many tutors as there were members of the church. Although group experiences might be even better than tutoring, this effort was a major statement of the church's faith and did change some lives.

Life-Changing Experiences: When I offered my services as a math graduate student to the neighborhood church's tutoring program, there seemed to be some consternation. I sat while the leaders consulted with each other. Soon the minister appeared and introduced himself. He looked at me in an evaluative, not unfriendly, way.

"So! You teach mathematics at the university!"

"Yes."

"I suppose you consider yourself good at teaching mathematics?" This seemed like an extremely odd question, but with considerable discomfiture, I decided to nod.

"Do you think you could teach anyone who really wants to learn mathematics?" I was already on my lifetime crusade. Of course, I thought I could! I nodded again, now very curious as to why I was getting this interrogation.

"If a student really wanted to learn, would you care what age the person is?"

"Oh, no!"

"I know someone who wants to learn mathematics very much, but she isn't a child any more. Do you think you could help her?" The purpose of the inquisition was becoming clearer. I assured him that age didn't matter.

"My wife wants to learn mathematics very much, but she is very frightened about it. Could you be kind to her, and not scare her?" I assured him I could. The challenge appealed. He looked me over and over.

"Okay. I'm going to get her. Please be as kind as you can be. She is very scared."

He then brought out a young woman probably a few years older than I was, who indeed was the picture of fright. After he introduced us, took us to an isolated room, and left us alone, I asked her what math she knew and what she wanted to learn. She looked at me, frightened and beseeching. She could count. She thought she could add. She wanted to learn whatever I could teach her. I swallowed. Her husband's concern began to make sense.

We began. She knew practically nothing, but learned quickly enough. Week after week, both of us came back. She learned most of arithmetic and some algebra in that one year. I learned how appallingly our system alienates some nice, eager people. I never learned where she went to school, or why she emerged so ignorant. It wasn't lack of ability or motivation.

That minister knew close up the misery that arises from lack of math. No wonder his church had a tutoring ministry! I assume he had prompted the leaders to watch out for someone like me. The effect on me was memorable.

Perhaps even more far-reaching was my own first grade teacher's putting me in the corner to help the two children who would be retained at the end of the year. It did more than keep the three of us from causing her trouble; it built friendships, and I began to see the satisfactions of teaching.

Robert Moses, an indefatigable leader in the 1960s Mississippi Student Nonviolent Coordinating Committee (SNCC) voter registration movement, tutored his own children twenty years later in Cambridge, Massachusetts, until the eldest insisted she wanted to include her classmates. The teacher was willing, so he began teaching math to the entire class. The results were sufficiently satisfying that he expanded his "Algebra Project" in northern cities.

Then he decided that, just as voting had been the road to freedom for Mississippi blacks thirty years before, now it was math. He returned to the same county where he had worked in the '60s. He and his colleagues teach algebra to all seventh and eighth graders in "their" districts, not just to so-called "gifted" children. He finds that sometimes children who had trouble with arithmetic excel at algebra. Moses believes all children are gifted if well taught—and when he teaches, they learn.[13,14]

The initial success of the Algebra Project in the Cambridge public schools inspired Freedom House in Boston to adapt it to an after-school program, which is now called Algebra Centered Enrichment (ACE). Both the school-based and the after-school program have yielded dramatic results. Parents in other places might follow Moses' lead, setting up another Algebra Project site or starting similar programs.

The Algebra Project runs parallel courses in algebra for parents. Robert Moses and his colleagues believe that mathematical literacy is needed for the postcomputer world in the same way that reading was needed for the industrial world.[15] Well taught, algebra can give parents an entirely different view of math than arithmetic and "can enable parents to advocate for change that will ensure that their children have access to the educational content they need to meet the emerging literacy requirements."[16] Obviously, parents who learn math together will be better prepared to collaborate on behalf of their children.

Your Own Initiatives

Like Robert Moses, other motivated parents can make a major contribution to their own child's schools if they are prepared and the schools are cooperative. If you aren't that ambitious, there are other ways to find or form a math community for your child.

Visit museums, libraries, and other community centers often with your child to see what displays and resources they have available. Take along one or more of your child's friends. Better yet, team up with other parents for regular outings to places that stimulate your children's interest in math and related topics.

Lead your child's troop temporarily for a special Girl Scout or Boy Scout badge. Look through the scout handbook to see what you can do. Consult with the leader, of course.

Seek out a gifted children's program at a nearby university or other institution. This suggestion bothers my wish to serve *all* children, but parents can rightly put their own children's needs first. (If enough parents seek alternatives, one can hope a democratic society will get the message.) I know of many youngsters in grades 4–8 who were saved from laziness and/or accelerating rebellion by a weekend program. After their craving to be challenged was recognized and satisfied once a week, they were able to tolerate the deadening boredom they faced Monday through Friday.

Hire a mathematician to teach a group of children on weekends or after school. This may keep your child intellectually alive if the school is offering only baby food. Abstract algebra, discrete math, statistics and probability, logic, and other fields can be learned by children (and certainly middle-schoolers) if a good teacher can be found. Graph theory, knot theory, and game theory are current research math fields accessible to preadolescents and their parents.[17] Inquire at your nearest university's mathematics department to find an interested part-time or full-time faculty member or a graduate student. Undergraduates and freelancers are other possibilities.

Accelerated tutoring of individual children doesn't work nearly as well. People need companionship to really excel in mathematics, so seek out companions, even if you think they aren't quite up to your child. Sharing the cost is nice too, but that's not the major point. You and your child need peers to sustain a long-term alternative to mediocrity.

Prod a local TV station to run a call-in program during which a talented math teacher answers math problems for callers, taking them thoughtfully through the various steps. Such programs can be very successful. They not only provide answers to specific problems but also demonstrate how to approach math successfully. A program in my own area designed for teenagers soon had young children and adults calling in. It's a good way to build math enthusiasm among the public.

Another "plus" is that a gifted math teacher who had expected to spend her life tucked quietly away with a teacher's salary is now a celebrity who is stopped on the street! Potential TV recognition may attract talented teachers who would otherwise overlook the profession.

Organize your own math club to provide opportunities for kids to enjoy math activities together. After the early grades, you need some background to carry this off successfully. But when your children are most vulnerable, you may be able to use the *Family Math*[18] book to lead your own successful program with other motivated families.

Try temporary home-schooling. This may seem like an odd suggestion for building community, but it had that effect for my firstborn. She told me toward the end of seventh grade that she wanted an opportunity to be free of school so she could really study hard for a change. With the help of the extended family (since I was working full-time), we managed to arrange for her to study at home with minimal supervision for the first four months of eighth grade. She went to school each day for the last period to take gym; the school said gym was necessary to comply with state law.[19]

We were all amazed (especially her) at how much better she fit in when she returned to school in January. Perhaps the others had matured, or perhaps her own ego had "centered," or both. In any case, she returned happier and more able to tolerate school's imperfections.

If other youngsters could take off a few months for lightly supervised study, would they too gain an increased appreciation of school and a more comfortable sense of self? Could a cluster of concerned parents provide a similar alternative for a small group? It might be cheaper than private school and less drastic than taking the entire responsibility of home-schooling upon oneself.

It is sad but true that most neighborhoods provide little support and considerable hostility for high math achievement. Finding like-minded peers is a great asset in giving your child the emotional strength to forge ahead in math. Your own good intentions will benefit from regular companionship with other parents who believe that math success is desirable and possible for their children. And grass-roots collaboration is pivotal in changing the culture so that all children thrive mathematically.

Structural Change for Your District and State: Mathematical Renaissance

We've got to break these chains before the system turns our children into slaves.
                    —*Polly Williams, Milwaukee education reformer*[1]

The restructuring . . . must go in two directions: toward increasing teachers' knowledge . . . and toward recognizing and using teachers' expertise. . . .

. . . in contrast with other countries that invest most of their education dollars in well-prepared and well-supported teachers, half of the educational dollars in the United States are spent on Staff and activities outside the classroom. . . .

If a caring, competent, and qualified teacher for every child is the most important ingredient in education reform, it should no longer be the one most frequently overlooked.

The education challenge facing the United States is not that its schools are not as good as they once were. It is that schools must help the vast majority of young people reach levels . . . once thought within the reach of only a few.

                    —*The National Commission on Teaching & America's Future*[2]

Change on the massive scale that is needed takes time. Yet pockets of excellence exist all across the U.S., and we have begun to implement meaningful reform.

                    —*Gail Burrill, president of NCTM*[3]

Private solutions to public problems have only limited success. We need systemic change.[4] Since our country has over 15,000 autonomous school districts, this requires 15,000 movements. If you are fortunate, you live in a district that is already changing. Otherwise, your district needs someone to initiate a reform movement in rhythm with the rest of the country.

Doing so makes one feel like a guerrilla—a rebel fighting the culture. Possibly, the most rebellious thing I ever did was to teach both my own children math when they were very young. When I empower, inspire, and prod other teachers, parents, and children to excel at mathematics, my opponents tell me how naughty this is. Yet, for humans to survive on this planet, we need many, many people who question, seek new answers, and are comfortable with mathematical thinking.

One person is a lonely voice in the wilderness. Two committed people can begin a movement. Surely you can find another dissatisfied person in your district. This chapter will show you where to look, and what you might do when you find kindred spirits.

The beautifully illustrated booklet *Everybody Counts: A Report to the Nation on the Future of Mathematics Education*[5] outlines our country's systemic problems from kindergarten through graduate school. Inexpensively priced for wide distribution, it discusses what various groups must do to attain national mathematical viability. Acquiring just one copy of *Everybody Counts* may validate your voice and will rescue you from feeling alone. Indeed, many U.S. citizens want an excellent education for our children; we need only find each other.

Unfortunately, there are also many U.S. citizens who are unduly content with their children's education. In Asia, both adults and children are far less satisfied, so they work harder.[6] The smugness that is so commonplace here is uniquely American. A former principal of Montclair's "International" magnet school (that has pupils from 35 different countries with 20 native languages) observed that not only were the visitors and immigrants from Asia and Europe academically way ahead of Montclair children, but those from

South America and Africa were also. Montclair has more than its share of educated parents, and the town is vastly wealthier than most of South America and Africa. The foreign students' math superiority over that of Montclair's children is a sad commentary on our country's priorities and policies.[7]

Elementary schools that hope to teach math well need to:

1. Upgrade the curriculum.

2. Hire and reeducate teachers so they know math and many ways of teaching it.

3. Revamp standardized testing or drop it.

4. Free teachers and principals to use their professional judgment.

5. Involve parents.

The first three steps, respectively, are addressed in three publications written collaboratively by the National Council of Teachers of Mathematics (NCTM) and published in 1989, 1991, and 1995. The first publication considers curriculum, the second teaching and teachers, and the third the evaluation of children. They provide coordination in a country named the *United* States. Although reading these national documents cover-to-cover is beyond the call of duty, people determined to promote change should become familiar with them. Their total cost is less than $100. As mentioned on page 237, you should be able to borrow them from your child's school—and also from your public library. The NCTM *Standards* documents serve as beacons, describing practical goals compatible with our culture, but capable of making us internationally competitive.

The first four steps are addressed prophetically in the 1996 report of the National Commission on Teaching & America's Future.[8] Successfully implementing the first four requires the fifth. James Comer[9] is a leader in showing how large-scale parental involvement can change schools, especially when students have backgrounds different from the staff. He recommends establishing a school governance team led by the principal and involving elected teachers and parents, a mental health specialist, and a member of the nonprofessional support staff. These teams (1) focus their efforts on solving problems (not blaming); (2) make decisions by consensus, not voting, since voting separates winners and losers; and (3) recognize the authority of the principal, but realize that other players are also essential for the success of a school. Many "Comer Schools" have improved dramatically by any reasonable standard.[10]

Presidents Bush and Clinton have tried to exert national leadership, but it is difficult in a country so devoted to states' rights and local control. Our federal government is a federation of states, and the states have retained most of the power over education. Each school district has ultimate power and responsibility. The route to reform is complex.

State Mathematics Coalitions

Since most educational regulations are by state, change is obviously needed at the state level. Toward this end, the National Science Foundation provided initial financing for a "Mathematics Coalition" in each state and the District of Columbia. The fifty-one Mathematics Coalitions have tried both to promote change and bring together kindred spirits within each state.

The National Alliance for State Mathematics and Science Coalitions[11] can direct you to your state coalition, if it still exists. Your state's Mathematics Coalition can suggest nearby friends and activities. If your Coalition sponsors a nearby Math Awareness Week event in April, try to attend. It's a good place to meet others with similar concerns. If your state's Math Coalition has died, the National Alliance may be able to give you the names of some of your state's math leaders; otherwise the organizations listed in Appendix B may be able to direct you to such leaders.

While working for statewide change, a district—even a school—can make great strides in improving opportunities for its own children. The video conference "Creating a Climate for Change: Math Leads the Way" has generated materials available to local change-makers.[12] It shows how inspiring schools have already made great leaps, and indicates that math, as a discipline, is already poised for change; other disciplines have greater difficulties.

The first three steps schools must implement to teach math well involve structural changes at the state level, but the fourth step requires only that states relax their stranglehold on the daily operations of schools. Although state officials ignore slight infringements of their rules when dynamic schools clearly improve the educational level of their pupils, principals need and deserve acknowledged power to serve the needs of their own children in collaboration with teachers and parents. Those who misuse this power need to be challenged (most likely by a coalition of parents and teachers) or replaced.

Who? Building Coalitions from the Ground Up

In a democracy, nobody has the power to dictate, so change requires cooperation among several groups. Change in the U.S. occurs only through coalitions, as French historian de Tocqueville noted in the early nineteenth century.[13] If planning change involves representatives from all affected groups, there is less resistance from a faction that feels excluded or misunderstood.

When forming a coalition to creat effective mathematical change, you need to involve people who care about excellence from most of the following groups. Fortunately, such people exist. The challenge is to find them! The joy when they first find each other is worth the effort.

Stereotypes have their risks, but a little commentary about each group may help you avoid some political bloopers. Of course, people within every group vary enormously. Generalites can be seriously misleading.

Mathematics educators: Members of the state NCTM affiliate should be familiar with the NCTM *Standards* documents and some will be willing and able to help implement them. Most high school and college faculties include some NCTM members. They may provide professional leadership.

Other mathematicians: There are about twenty-five professional organizations whose members took an undergraduate major in mathematics. Many have done postgraduate work outside education, concentrating on mathematical content. Most of them agree with the spirit of the NCTM *Standards* and could contribute expertise different from that of mathematics educators. They may know more about mathematics but less about pedagogy. On the whole, these two groups work well together. Both feel misunderstood by the general public and welcome the understanding of the other.

Elementary school teachers have been mathematically cheated by our culture. I have found them intelligent and eager to learn mathematics and full of wisdom in other matters.

Since there is a wide culture gap between them and mathematicians, interpreters may be needed. However, elementary school teachers and mathematicians have a great deal to give to each other. The pedagogical insight of elementary school teachers can be very helpful to people teaching mathematics at any level.[14]

High school teachers: The hostility between elementary and high school teachers is considerably greater than that between elementary school teachers and mathematicians. Many high school teachers were required to learn inordinately difficult math, only to end up spending much of their time teaching students who don't know third grade material by international standards. They have good reason to resent the *system,* but sometimes they vent their anger on the elementary school teachers who failed to prepare their charges for real high school mathematics.

Middle school teachers are a varied lot, without a culture of their own. Their members are drawn from the preceding two groups; often they share the problems of both. Elementary school teachers teaching in grades 6–8 tend to be overwhelmed by the subject matter and don't like being criticized for coping with what they consider (with good reason!) to be an impossible task. High school teachers teaching grades 6–8 are horrified by what they see their colleagues doing and upset by the needless damage inflicted on their students before grade six.

Some of the best teachers in the world may teach at this level, but also some of the worst. Many students are at their most difficult.

Parents run the gamut in mathematical preparation but have a strong vested interest in protecting their own offspring from being ruined by a system that has devastated so many others. They want change *soon,* in time to benefit their loved ones.

Principals are crucial in fostering or thwarting classroom change. Those with a math or science background are baffled and frustrated by what they observe in their schools. One actually taught three math classes a day in his elementary school, believing that while he could delegate other responsibilities, there was nobody else in the building who could empower the children to learn math. At the other extreme are those who believe that Drill and Kill is mathematics and are threatened by intimations that math education in their building is not what it should be.

Most principals fall between these extremes. They care deeply about the children under their care and the reputation of their school (two very different but valid motivations). They are genuinely uneasy about signs of mathematical trouble they perceive. Not being sure where to turn, they welcome collaborative help. They are beleaguered by a wide variety of problems that divert their attention from mathematics (and, alas, education), but they want to be cooperative. Getting their attention can be critical.

Curriculum Directors, Basic Skills Administrators, English-As-a-Second-Language Administrators, Special Projects Leaders, and other administrators display the same wide variety as school principals. One such person can make a huge different in effecting change.

Mathematics Directors have been neglected and undersupported, but where they exist, they are key. It is important to recognize the validity of their past accomplishments with limited resources and to work with them. All math directors should know and support the NCTM *Standards,* but their personalities and political problems vary.

If your school is dragging its mathematical feet, make an appointment with the mathematics supervisor of your school or district, if one exists. If not, it's a bad sign. Perhaps you can consult with the mathematics chair of your local high school or your Family Math teacher. With luck, you will find a kindred spirit in a position of authority with whom you can devise a collaborative plan.

Science teachers, directors, and specialists are often relieved to see signs of mathematical reform because it is extremely difficult to teach science to mathematically incompetent students. They are natural allies, but may perceive math as purely a servant and miss its full glory. A few, alas, may believe that a good servant doesn't think.

School board members are critical. They can ask questions, insist upon the formation of committees to develop plans, locate other "voices in the wilderness," and provide incentives and rewards for higher standards. They may be able to find funding, either by convincing taxpayers that math achievement is valuable, or by obtaining grants from private sources, or both.

Citizens knowledgeable in math and science are badly needed on planning committees. If they have time to volunteer in classrooms as teachers and/or tutors until more support is available for adequate math instruction, so much the better. Retired people and full-time parents (or parent substitutes) of school-aged children are two potential sources.

Funding sources ideally will be represented early in planning discussions so your collective dreams do not founder for lack of money.

The press is indispensable. It can present the challenges convincingly and mobilize the public. Imaginative education will help maintain a free press, as most reporters realize.

How? Educating Our Elementary School Teachers in Math

Providing elementary school teachers with a better mathematical education is essential for mathematical excellence. Our country needs more and better math education in both pre-service and in-service programs. ("Preservice" education is for people who hope to become teachers in the future, and "in-service" education is for current teachers.) Since most people who will teach your child are already teachers, you may be especially interested in in-service education.[15]

In-service education needs at least two components: classes for the teachers and demonstrations in the teachers' own classrooms. No matter how hard a teacher tries to change her style, it's almost impossible until she sees someone else with her own pupils. That raises consciousness!

Furthermore, children love having math visitors, and they enjoy the idea that adults are learning too. Once, when I was being evaluated, some third graders saw their teacher, their teacher's teacher, and their teacher's teacher's teacher all in the same room at the same time. Fun!

The "in-services," as they're often called, can be either classes at a nearby institution of higher education or independent programs (often called "workshops.") They can take place after school, on weekends, and/or in the summer. Lunchtime discussions have been used. Residential summer programs are probably the most effective. To provide time for regular continuing education, one teacher suggested an extra paid month in the school year (August?) for teachers only. Some districts put well-prepared substitutes in the teachers' regular classrooms so the teachers can participate in significant in-service programs during their own school year.

These programs cost money, and teachers should not have to pay. Reimbursement for tuition should be routine, and payment for study time should be considered. It takes *time* for anyone to learn mathematics, and the psychic baggage of many elementary school teachers often makes it emotionally traumatic as well.

Who should teach math courses to elementary school teachers? A coalition is best—preferably including a (collegiate, university, or industrial) mathematician, a math educator, a high school math teacher, and a math-enthusiastic elementary school teacher. The viewpoints of these four groups are remarkably different, but each is valuable in enriching the math perspectives of elementary school teachers. Members of these groups are compatible; there need be no conflict. Such groups, of course, can handle more than one class while they collaborate.

Most teachers welcome good in-service education and rapidly become better at teaching math. Those who are too math-phobic could be teamed with another teacher.

I believe that the fastest way to see dramatic mathematical improvement in this country would be to require every tenured teacher to sign a statement saying that she or he loves mathematics and wants to teach it. Those reluctant to sign such a statement should teach only subjects at which they feel competent. If there are really not enough enthusiastic math teachers available, then some children should go without "math" instruction for a year or two. It would be better for them than anti-mathematics taught by a math-phobic teacher.[16]

Pre-service education merits some attention, even if it's just your grandchildren's education and retirement security that are affected. They matter too. Eventually, your child may well get at least one new teacher, since we will be replacing half of our current teaching staff in the next decade.

People who have no idea of what mathematics is have much power in education circles. Indeed, some of these "educrats" have said to me, "I have become a [full professor, for example] without knowing how to find the area of a rectangle. Why should you expect it of third graders?" Children's math miseducation flows from theirs', and our country's economic competitiveness plunges. Your retirement security is jeopardized.

Who should decide what math your child's teacher knows? Currently, it's a political football. Mathematics and education professors see the priorities very differently, so teacher certification is, at best, a tug of war. Currently, mathematicians have "lost;" the educrats are both more numerous and better organized.

Harriet Tyson spends a chapter in *Who Will Teach the Children?*[17] describing teacher certification methods in detail. Many professions, from physicians to cosmetologists, monitor their own profession via nationwide organizations, but teachers are ostensibly licensed by and for each of the 50 states. Tyson in *Who Will Teach the Children?* and Rita Kramer in *Ed School Follies: The Miseducation of Teachers*[18] paint a bleak picture of what actually happens in our teacher preparation institutions.

One problem is that states can fine districts for not providing adequate education, even after requiring districts to hire from teachers certified by that particular state and not providing adequate preparation for its "certified" teachers. So there is a high incentive for districts to deceive the state, and, therefore, the public. "A hospital lacking a qualified doctor would be required to close down. . . . An unlicensed electrician is not allowed to work on your household electrical system . . . [but] schools cannot be closed down for lack of licensed teachers. . . . When faced with shortages, states quietly lower the cut scores on their tests." . . .[19]

Another problem is that certification by each state means that good teachers can't easily move among the states. An oversupply of teachers in one state can't go to other states that need them, as happens in most careers. Tyson hopes for national credentialing by the National Coalition for the Advancement of Teacher Education (NCATE), similar to the professional credentialing in many other careers. If we are to achieve academic excellence, however, subject matter specialists absolutely must have a "say" in determining teachers' competence in their field. Political problems continue; only active collaboration among educators, mathematicians, and the public will raise standards.

Whatever the credentialing requirements, we need to encourage people who want to devote their lives to teaching children to take as much mathematics in college and graduate school as they like. Having so many other required courses that prospective elementary school teachers don't "have time" to become proficient at math and/or science is suicidal for a modern culture.

What Else?

Many other changes, along with improved math education for teachers, are needed for our schools to provide a consistently excellent math education.

Schedule in collegiality among teachers. Teachers should have time to collaborate with each other. Better than anyone else, they can help devise ways for children in that school to learn. In other countries and at the collegiate level, such collegiality is deemed essential.

Schedule more time for mathematics. Political pressures are enormous for teachers to neglect mathematics, even if it is their favorite subject. Since many educational administrators hate math, serious political pressure is needed within educational circles to challenge anti-math bias. Hirsch's writing[20] (among others) emphasizes that in countries where math receives more time, achievement in all subjects rises.

Drastically upgrade the curriculum in many math subjects, but two merit special mention.

Statistics should be taught in every grade, beginning in kindergarten. A public that doesn't understand graphs and data presentation can have much wool pulled over its collective eyes. If we are not to sink into an economic morass, we need massive remedial teaching of modern statistics *soon* to many U.S. adults—not just to elementary school teachers.

In other countries *calculus* (the mathematical analysis of change) is required for college admission. The good news is that the percentage of U.S. high school students taking calculus almost doubled (from 5% to 10%) in the decade 1982–92.[21] You and I are part of a movement! On the other hand, in 1978–79, studies indicated that over ten times as many students in the Soviet Union were studying calculus as in the U.S., although their total population was only slighly more than ours.[22]

Furthermore, (as noted on p. 3) international tests in the mid-1980s found the *top* 1% of U.S. twelfth graders at the *bottom* of the top 1% of the other eleven countries studied.[23] In other words, our very best seniors tested below the very best of these eleven other countries. Our top 5% rated below the median (average) scores of Japanese students in twelfth grade tests.[24]

Inability to analyze change (the subject of calculus) seriously diminishes the capability of politicians, religious leaders, media people, and others who can "turn" the American tide.

Schedule more recesses.   Children could concentrate more if they had more breaks for exercise and informal socializing.

Teachers need planning and collaborating time, both during the children's recesses and during weeks set aside (in August?) for this purpose.

Give math more in-service time and give teachers financial support to attend math conferences and courses on their own time.   Many more professional educators are reading specialists than math specialists. I have no quarrel with their claim that reading is even more important than math, but considerable evidence shows that math power contributes to reading ability. Furthermore, every elementary school teacher I have met can read just fine, thank you. (Many teachers need remedial work in writing, but not as many as need it in mathematics.)

Provide mathematics specialists in elementary schools.   Until classroom teachers are better equipped, each school needs at least one type of mathematics specialist, and probably two. One teacher in each building should be comfortable and competent with mathematics and be a daily resource for the others. Also, a serious math-lover from "outside" should visit each school on a weekly or biweekly basis to address more sophisticated problems. This person could be a high school or college teacher, an industrial mathematician or engineer, or a retired person. Each elementary school teacher should have regular demonstration lessons in her own class by someone with no evaluative power over the teacher, so that the teacher and visitor are free to converse openly about the mathematical progress of the children.

This is not a novel idea. Already by 1971, the United Nations Educational, Scientific, and Cultural Organization was advocating that worldwide, "each primary school should include at least one teacher with special qualifications in mathematics who can act as an advisor." Earlier, in 1960, "the Mathematical Association of America strongly urged that at least 20 percent of primary teachers in each school should have a more advanced mathematical training comparable with that required in junior secondary schools."[25]

Have specialists teach math classes after third grade.   Some schools accomplish this by having two teachers share two classes. One teaches math, science, and health; the other teaches the other academic subjects. Personal understanding can be deepened if they keep the same children for two or three years.

Encourage math buffs to become elementary school teachers. The current custom of discouraging mathematically competent people from entering "elementary ed" *must be stopped.* We need as many math enthusiasts in elementary education as we can get. Too many adults have been told, "You're too good at mathematics to become an elementary school teacher. It would be a waste." It is *not* a waste to provide little children with good math!

Mobilize parents. Schools can't do everything. More effort is needed to prepare families to think together. Parents, like teachers, need supportive opportunities to learn and enjoy mathematics, so they can overcome inadequacies in their own backgrounds and enthusiastically foster their child's math education. They need to experience joyful math so they can better guide their children.[26]

Where to Cut?

If we put more emphasis on real math, what goes? Many topics can be integrated with math, so an able teacher can teach both math and something else at the same time. Quality goes up without eliminating anything.

However, integrating subjects has its limits. Some activities must be de-emphasized to include more challenging subjects. The following are some places to cut (or cut back):

Drill and Kill, certainly. Kill it! This would save a great deal of time.

Most testing, especially multiple choice tests. In elementary schools testing should be either revamped or dropped. If we can't afford to do it right, we can't afford it. Our current testing habits are unfair to females and minorities and don't do any favors for white males either.

Written standardized testing in K–2 should cease altogether. Young children need to be loved and nourished. Caring, competent teachers need to continually assess their pupils' progress by *watching* them.[27]

Repetitive remedial programs rarely work. According to most evidence, traditional remedial "learning" rarely lasts beyond the test to which it teaches. Teachers must teach mathematics correctly in the context of the regular classroom. To do this the teachers must know and like the subject.

Gifted children programs now provide opportunities for a fortunate few. They should be available to all but the *very* few who cannot master a challenging curriculum. Parents must demand a challenging curriculum for *all* children.

Other "pull out" programs should be reconsidered. Flexibility is needed within the regular classroom. Well-prepared, well-supported teachers who are teaching small classes can provide individualized instruction. They know best what each child needs by watching every day. Eliminating pull-out programs may or may not cost jobs; the special ed teachers might help regular teachers within the classroom or they could teach their own regular classes.

Forced "mathematics" teaching doesn't work. Teachers who are teaching so-called mathematics only because they are forced to do so are wreaking untold damage on U.S. children. Elementary school teachers who know mathematics want to teach it. No teacher should be *allowed* to teach mathematics unless he or she wants to.

The more I see of U.S. mathematics education reform, the more rigid I become about one goal: flexibility. We have so far to go! The 15,000 school districts are so different! The teachers within each district are so different! The children, the families, the parents, are so different! The only viable approach is to listen to each other, and, working *together,* we can begin to effect change.

Will the Real Mathematics
Stand Up and Be Recognized?

22

There are three kinds of mathematicians:
those who can count and those who can't.

> —*Joke circulating in the U.S. mathematics community, 1990s*

When the president of the Mathematical Association of America began a meeting of the Board of Governors with this joke, laughter of recognition broke out throughout the room. Mathematics is not always what it seems to the general public. How did such a fascinating pursuit as mathematics get warped into a school subject that so many students detest? The reasons are complex, and the solutions aren't simple.

Humans were created to enjoy thinking. The natural "high" of suddenly comprehending a new concept is unsurpassed in its magic. Children whose parents preserve their natural joy in these "highs" will be less likely to cower before giants—whether those giants are people, drugs, or ideas. The native abilities of these children will blossom.

Our math "ability" depends on how hard we work and the teaching we get. The best teaching is subtle and motivates students to work hard. These statements are supported by my study of twenty-one of the first twenty-four U.S. black women to earn the highest possible degree in mathematics—the Ph.D.[1] They all worked very hard and struggled against formidable odds. They also reported two other characteristics in common:

1. All could remember a secondary school teacher who had told her that she was gifted at mathematics and that it would be worth her while to struggle for high math achievement.

2. All had at least one relative in the previous generation (usually a parent) who was willing to sacrifice significantly for her education.

Many male and white mathematicians also had these advantages, but not all.

Why are so few young people lucky enough to enjoy mathematics? Pervasive myths that too often mask real mathematics do incalculable damage. In this chapter, we'll review some of the most damaging myths this book has already addressed, concluding with three more.

Damaging Myths

Mathematics ability is innate. If your math power is predestined, why work at it? Why should teachers help the "stupid" kids? Listening to elementary school teachers has convinced me I'm a math lover because of the way I was taught by my parents and teachers. Working with children has convinced me that my parenting, not my genes, caused my children to do math better than I did. Others also learn math fast when I entice them to do so.

Hard work is unpleasant. Hard work can be great fun. Dedication to an important quest larger than oneself can lead to the greatest joys of human life. The adage "no pain, no gain" has been overdone. Some "gains" in love, reading, tennis, and math can be attained pleasantly. Nothing is always pleasant.

Children naturally resist learning. Just the opposite! All children love learning, including math, until they are taught to hate it.

Children's minds are empty slates on which we learned folk write. Children's minds are more than that; don't ask me what.

Mathematics is just cold, clear logic. Logic is essential, but not enough. Pure logic has its limits, even in mathematics. Intuition plays a major role too. Beauty and diversity must be included in effective mathematics education.

The fixed hierarchy of mathematics means the "basics" must be "mastered" before more difficult topics. Often learning a higher form of math sheds light on the easier topics. I am repeatedly impressed by how much I learn each time I "teach" a first grade class; these are hard concepts, and I'm still learning about them. (However, expecting children to master place value before knowing multiplication destroys the self-esteem of many U.S. children.)

Drill and Kill is mathematics. Real mathematics concerns patterns, not facts. Children who are excited about real math memorize the necessary math facts without misery.

Mathematics is a fixed body of knowledge. The ancients did mathematics, and much of their math is still useful. But human knowledge of math is growing so fast that some leading mathematicians refer to the late twentieth century as having a "mathematical explosion." [2]

Girls who are good at math have trouble attracting men. The supply is less than the demand, so the selection of men for math-loving women is greater than the selection for most women. Furthermore, the quality of men available to math-smart girls is better than average. Intelligence is like a skunk spray against less desirable males.

Smart people are (nearly?) crazy. Many smart people are also delightful, happy, generous, open-minded, and useful. The most notable discovery in my surveys of mathematicians is how satisfying their lives are. The widespread fear of becoming smart is sad and misguided.

Unfortunately, math hostility is rampant.

June Porto used to be uncomfortable with mathematics but learned to enjoy the mathematics she was teaching to elementary school children. When she was appointed to the Math Lab of her school to teach mathematics full time to children who were having trouble with it, she was startled to discover how much her reputation was affected. "Oooh!" people would say drawing back in distaste. "You teach mathematics! Ugh!"

Teaching your child that it's okay to love math may be the most important way you can help your child achieve. If you don't have a dream, you can't have a dream come true. But many U.S. youngsters *don't* dream and learn because they fear social disdain.

Most teachers of all subjects and at all levels feel like we're in sales—always pushing the joys of the intellect. Some years ago, two surveys were done at Montclair State at about the same time. When the faculty were asked about their primary goals, 80% said "Teaching students to think." However, when students were asked what made a faculty member excellent, only 5% even had "teaching us to think" on their list.

Most of my students seem very grateful when I stimulate their thinking. Alas, they don't expect an opportunity in the classroom. How sad! Obviously, my colleagues are trying. Judging by many conversations with teachers at all levels, most are. Why don't U.S. kids believe it's okay to think in school? Why do so many think working hard isn't fun, although they once obviously enjoyed it very much? Why has sloth become so appealing and socially acceptable?

We Can't Know Everything: Gödel's Incompleteness Theorem

If you're feeling somewhat overwhelmed by the sheer amount of mathematics out there, don't take it personally. There's more mathematics than all humankind can learn—even if we work at it forever. This has been proved—mathematically!

In 1930 Kurt Gödel proved that we can *never* know all the mathematics there is. If statements can't be both true and false ("inconsistencies," in math jargon), there will always be more math. (If we allow *some* math statements to be both true and false, then *everything* is both true and false—but this is not a satisfying state of affairs.) So no matter how much we, as a human species, discover (or invent) about math, there is always more.

Worse, Gödel proved there are some specific questions that we will never be able to answer. He correctly guessed one of them: the so-called Continuum Hypothesis. Since then, people have proved that indeed the Continuum Hypothesis cannot be proved or disproved. Now there are two "strands" of mathematics, one assuming the Continuum Hypothesis to

---

Inconsistencies Are Unavoidable

Douglas Hofstadter's best-selling book *Gödel, Escher, Bach*[3] is a popular exposition of Gödel's ideas, comparing them with those of a famous artist and composer. Snippets of *Alice in Wonderland,* other writings of Lewis Carroll, and Hofstadter's own fantasies in Lewis Carroll's style are interwoven among the mathematical and philosophical discussion.

Gödel's Theorem is related to this ancient paradox: This statement is false.

If this statement is true, then it's false. If it's false, then it's true! Apparently, logic is not infallible. Sometimes it plays us false. What does this imply for the rest of our life?

be true and the other assuming it to be false. We will never be able to use math to tell whether this mathematical statement is true or false!

The recognition that people will never know some things in mathematics has far-reaching implications. Apparently, "undecidability" is part of modern life. We must resign ourselves to fundamental ignorance. We need not feel embarrassed to say, "I don't know" to our children, no matter how young they are. We (and they) need to keep learning because nobody knows all the answers.

As we study mathematics, we become increasingly aware of its limits. Like any good friend, mathematics loses the luster of apparent perfection when we know it well. Like other good friends, we may worship it less but love it more as we become closer. Studying mathematics profoundly affects our world view.

Twentieth-Century Math: More Than You Thought

Arithmetic is only a small part of mathematics. Unfortunately, in the 1960s, in an attempt to broaden the elementary school curriculum, arithmetic was renamed "math." This was unfortunate, because math also includes such diverse topics as logic, geometry, statistics, planning, and calculus. The curriculum still needs to be broadened; renaming is not enough.

If we believe that humans are designed to enjoy hard work (as toddlers clearly do), one way we can entice Americans to engage with math is to offer more challenging math topics in the schools. Underachieving youngsters often fulfill high expectations when able teachers offer challenging math. Dumbing down is not the answer.

Calculus (the study of change) has been one casualty of our country's acceptance of sloth. Since change is the one constant in today's world, other countries require calculus for all high school students hoping to attend college. Thus, their prospective elementary school teachers study math through calculus before *entering* their teacher preparation programs. Until our country provides comparable preparation for those who influence our children's math attitudes and backgrounds, we must provide plenty of remedial math for elementary school teachers, during and after college.

> **Miracles in Downtown Newark:** A third grade teacher in a difficult, drug-ridden neighborhood of Newark, New Jersey, greeted me one day with the question, "Would you be willing to set aside our lesson plans and answer the children's questions that I can't answer?" Intrigued, of course, I agreed.
>
> The next hour was one of the most exciting of my life. The best students asked the questions, but all were able to follow the answers. They and the teacher sat on the edge of their seats, paying close attention. I lost myself in the wonder of it.
>
> "What do you call this kind of mathematics?" the teacher asked after we emerged at the end.
>
> I gasped. "Why, this is the beginning of calculus! This is a lesson I teach to college students at

the beginning of a calculus course." Although I had to teach division incidentally along the way using Cuisenaire rods, the children had been able to follow the concepts of limit and infinity, rarely taught before college in our country.

A few years later I had a similar experience with another group of impoverished urban third graders. This time I realized what I was doing, but the lesson was similar. A few children asked questions leading to introductory calculus, and I let the discussion follow their curiosity. In both cases, all children in the room seemed to enjoy the adventure, gliding together into the realm of higher math.

Such teaching brings me unspeakable joy. To help people catch a "glimpse of the beyond" is to share a moment of ecstasy. Some elementary school teachers have noted that, rightly taught, mathematics is one of the easiest subjects to teach. The rewards are intrinsic; there is no need for external rewards or applications.

Making Connections

Sometimes, of course, we like to make connections to other subjects. However, when we do, we are also studying cooking, carpentry, physics, or economics—something other than just math. That is fine. However, by itself, math offers an uncomplicated world, free from human desires and evils, where there are *right* answers. Only in math are there answers that everyone on Earth agrees are right. Any other subject stirs up discord, where someone may say, "Yes, but. . . ." Children, like other people, enjoy the comfort of knowing they are absolutely right. I have seen elementary school classes choose abstract math over math connected to applications. There is a magic in pure math that unruined children appreciate.

Racism is no excuse for our country's math inferiority. We come from the same gene pool as people in China, Great Britain, Italy, and Nigeria. If they can learn challenging math—and they do—so can our youngsters. It's time for U.S. society to change its attitude toward intellectual achievement, and parents have the first opportunity.[4]

Academics, including research mathematicians, must be included in discussions about reforming math education. Dismissing the ivory tower as "impractical" reflects the same wasteful attitude as indulging adolescents in material goods without challenging them intellectually or taking their opinions seriously. If intellectuals are not part of the conversation about education, our country may lose its intellectual heritage. If we academics have been too cantankerous, we too must be invited to change. But people are less cantankerous when they feel someone is listening. This includes both intellectuals and adolescents.

There are fewer U.S. citizens holding a Ph.D. in mathematics in the entire country than there are lawyers in New Jersey. A Ph.D. in any subject requires original research in that field, and doing research in mathematics means thinking up a theorem that nobody ever knew before and proving it. It's hard! It doesn't become easy (just more fun) because one loves math. One can lose much time trying to prove an untrue theorem with nothing to show for it. Unlike natural science, where an experiment to show something is true may instead yield publishable results that it is *not* true, running down a mathematical dead end rarely gets into the journals.

When university administrators insist that mathematicians churn out new theorems (for publication) with the same regularity that scientists do laboratory experiments or historians examine documents, it's no wonder that the subject has fallen upon hard times. Mathematical research is more like creating a symphony than studying music, as most music professors do. Across the country math departments are withering, gutted by administrators with no understanding of mathematics. In defense of administrators, however, I should note that their ignorance is symptomatic of a much larger problem. "Welcome Mathematics Teachers," said a banner at an event for the 1986 International Congress of Mathematicians when the group met in the United States. I cringed for my country's ignorance about the breadth of what mathematicians do.

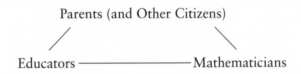

Parents (and Other Citizens)

Educators ——————————— Mathematicians

The mathematical education of children is rightly the joint responsibility of parents (and others who take a personal interest in children), educators, and mathematicians. The three groups must collaborate if sound decisions are to be made. Allowing professional educators to determine what teachers and children learn is like allowing nurses to determine what doctors are taught in medical school. Yes, nurses and teachers deliver invaluable, underpaid services that our lives depend upon. Because they are so overworked and underpaid, however, they rarely have time to scan the mountain tops.

Our schools must reconnect with those who have the time and background to contemplate intellectual vistas. Mathematicians, educators, and parents must converse together. Our children's future depends upon it.[5]

MATH FOR THE FUTURE

This country needs new dreams. It needs some drastic changes, as almost everyone will admit, and change isn't easy. In particular, we need alternatives to our cultural veneration of physical objects. One alternative is math. Math can be done alone, or it can be a pleasant community activity for family and friends. It can also be a source of spirituality.

Many mathematicians are deeply religious people, and our religions span the entire globe's options.

Our culture almost worships frenetic activity. People brag about how many hours they worked and how much sleep they lost. Are we solving the right problem? So much effort seems to go into solving problems that are of dubious value. We need more often to ask "Why do you think that?" and "Is there another way?"

It is also true that math broadens individual career choices and is vital for our country's economic competitiveness. A recent survey indicates that many academic departments welcome undergraduate math majors into their graduate programs, including botany, business, city and regional planning, computer science, dentistry, electrical engineering, entymology, geology and geophysics, government and politics, industrial engineering and operations research, journalism, law, linguistics, medicine, microbiology, music, nuclear engineering, optometry, philosophy, policy studies, psychology, and zoology.[6] If your child becomes enthusiastic enough about math to pursue it in college, there are many options afterward. For the rest of us, the above list indicates the pervasiveness of math in meeting the problems of today's world.

Without a critical mass of math power in the most powerful country in the world, we may fail, as a human species, to cope with the challenges our technology has spawned. The more of us that survive and thrive, the more complex our problems, and the more approaches we'll need to solve them. We neglect math at our peril.

Appendices

Where do I go from here? This book is written to be self-sufficient for parents whose children are in satisfactory schools and who do not have the time or reason to organize for better. Even they will be interested in books to share with their children. I hope they will also consider math books to enrich their own understanding of this potentially delightful (but too often misunderstood) subject. The first two annotated bibliography sections serve these needs.

Parents who are restless, perhaps harboring a wish to improve mathematics education for more children than just their own, should be aware that they don't have to start from scratch. There are national policy documents (Appendix A) and organizations (Appendix B) that can help them.

The charter school movement may appeal to such parents. It enables American parents (as well as teachers and other citizens) to bond together, preferably with a profit-making corporation or a non-profit that will help them raise supplemental funds, and apply to the public school system for an alternative school partially financed by taxpayers. Charter schools are distinctive schools within the public system; they must be tuition-free. Starting a school is a tremendous amount of work, but some families have found it well worth the time. (I recently attended the 35th celebration of the cooperative nursery school I started for my own children, and I'm sure I was the happiest person in that crowded room.)

Over a million American children are now being homeschooled, and the number is growing all the time. The most frequent request I had from readers of *Math Power* was for suggested curricula, texts, and other aids for homeschoolers. Thus, Appendix C provides resources for homeschooling parents. Some of these may be useful for other parents.

In composing Appendix C, I have been extremely fortunate to have the able assistance of Susan L. Schaeffer, a mathematics advisor to homeschooling parents in North Carolina. Thank you, Susan! I am grateful, and I'm sure many readers will be too. Recommendations that are entirely hers are marked "(SLS)."

Appendix A: National Policy Documents

Principles and Standards for School Mathematics, National Council of Teachers of Mathematics, Reston, VA, 2001, www.nctm.org/standards. This seminal, wise, and comprehensive document subsumes three from the previous decade and is a readable account of the direction mathematics education should be taking. It clarifies some misunderstandings of the previous documents and elaborates on the basis of experience with them. It should be in your school's library, available for parents, faculty, and administrators. If not, ask your school to acquire it. Meanwhile, you can read it free for 90 days on the web, or indefinitely if you join the NCTM.

Mathematics and Democracy: The Case for Quantitative Literacy, Lynn Arthur Steen, editor, Woodrow Wilson National Fellowship Foundation, Princeton, 2001, is a succinct and articulate statement of why all Americans should learn mathematical and statistical approaches to thinking—and some skills. Its message shows why teaching children young, when they learn most easily, is crucial. You can download it for free, one chapter at a time, from www.maa.org/mathanddemocracy.html.

Before It's Too Late: A Report to the Nation from The National Commission on Mathematics and Science Teaching for the 21ˢᵗ Century, 2000. A blue-ribbon commission appointed by President Clinton and led by astronaut John Glenn makes three strong recommendations for improving mathematics and science education. Available free from www.ed.gov/Americacounts/glenn or 877-4ED-PUBS or 800-USA-LEARN.

Everybody Counts: A Report to the Nation on the Future of Mathematics Education, Mathematical Sciences Education Board, National Academic Press, 1989; 2101 Constitution Avenue, NW, Washington DC 20418, 800-625-6242, 202-334-3313. Summarized the basic issues facing our country in mathematics education. An "oldie but goodie." Now available free online at www.nap.edu/catalog/1199.html. Hardcopy still available for $8.95, and well worth giving to principals and school boards.

There are other excellent documents from the same source. Two are:

Measuring Up, 1993, www.nap.edu/catalog/2071.html. Thirteen questions that test fourth graders' creativity in mathematics.

Reshaping School Mathematics, 1990, www.nap.edu/catalog/1498.html. Focuses on essential ideas that transcend the details of current curriculum and assessment discussions.

Schooling, Statistics, and Poverty: Can We Measure School Improvement?, Stephen W. Raudenbush, Educational Testing Service, Princeton, 2005. A provocative example of contemporary applied mathematics. What can be measured? How?

The Third International Mathematics and Science Study, TIMSS, collected data on mathematics and science education by giving examinations in fourth, eighth, and twelfth grades to a half million carefully sampled students in 41 countries. Many videos of classrooms were also taken. There is an abundance of books and videotapes based on the plethora of evidence mined from this study. Publishers include Kluwer Academic Publishers, Boston; the International Study Center Lynch School of Education at Boston College (ICA); Mid-Atlantic Eisenhower Consortium for Mathematics and Science Education, Philadelphia, PA; and the National Institute for Achievement, ᶜ ᵃrriculum, and Assessment, Office of Educational Research and Development, U.S. Depart ˌ ˥t of Education, Washington, DC.

Appendix B: Organizations

National Council of Teachers of Mathematics (NCTM), 1906 Association Drive, Reston, VA 22091-1593, 800-235-7566 for ordering materials, 703-620-9840 otherwise, www.nctm.org. Membership is open to anyone and is appropriate for a parent who wants to become deeply involved in mathematics education.

Benjamin Banneker Association, P.O. Box 24182, Lansing, MI, www.bannekermath.org, director@bannekermath. Affiliated with NCTM. "Dedicated to mathematics education advocacy, establishing a presence for leadership, and professional development to support teachers in leveling the playing field for mathematics learning of the highest quality for African American Students."

Eisenhower National Clearinghouse of Mathematics and Science, 555 New Jersey Avenue NW, Washington, DC 20208, www.enc.org. Produces many worthwhile items.

National Association for the Education of Young Children, 1509 16th St. NW, Washington, DC 20036, 800-424-2460, www.naeyc.org, webmaster@naeyc.org. Promoting excellence in early childhood education. If you search on "mathematics" at its Web site, there are many possibilities.

Women and Mathematics Education (WME), www.wme-usa.org/pages/364182/index.htm. Affiliated with NCTM. Designed for teachers who are concerned about gender equity, but open to all for a $15 per year membership fee.

Appendix C: For Homeschoolers

Suggestions specifically from Susan Schaeffer are marked (SLS). She can be reached at oaktreemath@earthlink.net.

THE BASIC CURRICULUM

Homeschoolers have varying reasons to homeschool and many philosophies of education. I am not an expert on choosing homeschooling materials, but after six years of hosting a radio show about mathematics, I have access to different types of experts in mathematics and mathematics education, and I have asked them for curriculum/text suggestions for home-schoolers. After considering their responses, I submit the following list. It is certainly not complete (there must be other fine programs), but it covers a breadth of approaches.

One major caveat when reading *any* mathematics book or Web site: *Do not believe that every sentence is true!* If it seems wrong, it may be. Talk it over with a friend or a more math-ematically experienced person. Everyone makes mistakes, even mathematical writers. Some Web sites have places where you can report mistakes or misprints in their materials. One excellent graduate math text had so many errors that I opened every class by asking what errors in the book the students had found. The basic ideas were right and well expressed, but you couldn't trust any individual sentence—or any answer to the exercises. It kept the students "on their toes." Keep yourself and your children alert and sufficiently confident to believe that you can be right when the book is wrong.

I am pleased that I have not had any reports of errors in *Math Power*. Julie Stillman was a wonderful editor.

The Singapore national series (www.singaporemath.com) provides all the basic elementary school mathematics presented correctly. It was written in and is used throughout that country. That Singapore often places first on international mathematics tests speaks well for their books, although an educational program involves far more than merely the written word. The series is written in English and has many avid fans in the United States. The individual books are environmentally responsible light paperbacks. It is available from "Family Things," listed later in this appendix.

Miquon Math, available from Key Curriculum Press, is a series of six workbooks for children in the early grades. They are mathematically rich, while being inexpensive and fairly simple for parents to use. They are a refreshing change from the usual drill-type workbooks; they focus on concept development and understanding of the ideas behind the basic operations, fractions, and geometry. The material is presented in an engaging manner with age-appropriate activities and many lessons that utilize the Cuisenaire rods. Susan Schaeffer reports, "When my own children worked through these workbooks, I allowed them to progress at their own pace; they saw this as a fun activity, and they liked to take charge of their own learning. It was excellent preparation for future independent work and gave them a great foundation in many basic ideas of mathematics. I credit Miquon with contributing to their lifelong enjoyment of mathematics." (SLS)

The "Key to" workbook series (i.e., *Key to Fractions*), also available from Key Curriculum Press, is a good sequel to Miquon Math. They are available in the following topics: fractions, decimals, percents, measurement (both English and metric), and algebra. They help establish a solid foundation in the mathematics of late elementary school and middle school, and they are easy for both parent and child to use and understand. (SLS)

Susan Schaeffer continues, "I often recommend Miquon and the "Key to" series to parents of homeschoolers as part of an eclectic curriculum. The rest can include traditional texts, problem-solving books, library books, games, Family Math activities, computer programs, videos, manipulatives such as Cuisenaire rods, geoboards, dice, blocks, household materials, and various other resources such as those listed in the appendices. These can be supplemented by study groups and math clubs, which might be organized around preparing for Mathcounts or the Math Olympiad for Elementary and Middle Schools. Homeschoolers can plan out their studies over the years. Programs such as Singapore or Miquon and the "Key to" workbooks provide structure and continuity, while the supplemental materials enhance the enjoyment and depth of mathematical study." (SLS)

If you want something more flexible that will challenge your own teaching abilities, you could try a program developed in this country in the past 15 years, two of which are mentioned next. (I cannot recommend the old-fashioned U.S. math texts; they were too narrow and too full of outright errors, not just typos and mistakes in answers.) These were written with the expectation that teachers would have lots of support from someone familiar with mathematics and mathematics education. Even the most enthusiastic supporters doubt that they are suitable for homeschoolers because of the lack of expected support. One person deeply involved in devising and publicizing them wrote, "It is clear to all of us who have been involved in developing a curriculum that any curriculum materials, no matter how well they can be used, can also be used badly, can be misunderstood and distorted." If you use recently developed American materials, you should either read prodigiously and widely or seek support. Some is available on the Internet. One educator suggests that parents who want to teach their children using a modern American program (perhaps because it is being used in their local school district) should also use a book such as John Van DeWalle's *Elementary and Middle School Mathematics: Teaching Developmentally*, which has already gone through at least six editions with varying publishers.

Two of the most popular new programs are:

- *Investigations of Number, Data, and Space*, TERC, 2067 Massachusetts Avenue, Cambridge, MA 02140, 617-547-0430, http://investigations.terc.edu. Publisher: Scott Foresman, 1900 E. Lake Ave., Glenview, IL 60025, 602-508-81218, investigations.scottforesman.com.
- *Everyday Mathematics*, the University of Chicago School Mathematics Project, (UCSMP), Everyday Mathematics Center, 5640 S. Ellis Ave., Box 15, Chicago, IL 60637, em-center@listhost.uchicago.edu. Publisher: Wright Group/McGraw-Hill, 1-888-772-4543, www.wrightgroup.com.

The Family Math books (www.lawrencehallofscience.org/equals) listed in the first section of the Annotated Bibliography may appeal to you if you prefer to avoid the sense of "lessons" or do not want your child to feel "structured" in the sense of any school curriculum you might consider. The director of the International Family Math Program tells me she believes that these books form a fine basis for an elementary school curriculum. I have not reviewed them for their mathematical completeness, but I *can* report that I have used them successfully to teach many concepts, and that children invariably delight in Family Math.

However, there are so many games that no family can use all of them, so it is important to choose at least one that teaches each of the basic math concepts. With the help of Chapters 11 and 12 in *Math Power* and the introductions to the chapters in *Family Math*, alert parents may be able to ensure that their children absorb all the basic ideas of elementary school mathematics—and a great many more as well.

Whether or not you use them as a text, the Family Math books provide enticing ways to learn the basic facts without deadly drill. Thus, I would recommend them as a supplement no matter what your curriculum.

www.homeschoolmath.net has lots of math resources, including reviews of a large number of math curricula that are designed for or commonly used by homeschoolers. The link to the reviews is at the bottom of the main page, on the left side. (SLS)

OTHER RESOURCES:

Child's Work, Nancy Wallace, Holt Associates, 1990. Has an excellent chapter on math and homeschoolers. (SLS)

Family Things, 19363 Willamette Dr. #237, West Linn, OR 97068, 503-727-5473, fax: 503-722-5671, www.singaporemath.com, customerservice@singaporemath.com. Carries the Singapore mathematics text series and some other books.

FUN Books, www.fun-books.com, an outstanding catalog for homeschoolers. There is an extensive math section along with other topics. There are standard books and an interesting set called "Murderous Math" on topics varying from number sense and fractions to probability and trigonometry. (SLS)

Home Education Magazine, Box 1083, Tonasket, WA 98855, www.home-ed-magazine.com. Many issues of this magazine contain reviews of math materials and articles about math. Lists of back issues in print and what major topics are in each issue are available in some issues of the magazine. The Web site has discussion boards and resource lists, among other things. (SLS)

A to Z Home's Cool, www.homeschooling.gomilpitas.com/materials/MathCurriculum.htm is the math part of the "A to Z's Home's Cool" Web site, which is a tremendous resource for homeschoolers in every area. (SLS)

The Home School Manual, for Parents Who Teach Their Own Children, 4th Edition, Theodore E. Wade, Jr., and others, Gazelle Publications, 1991. Has a thorough discussion of math—excellent reading for homeschooling parents. (SLS)

Homeschool Resource Guide, members.cox.net/ct-homeschool/guide.htm. Addresses and other information for vendors who carry math curricula and other math materials. (SLS)

Learning All the Time, John Holt, Addison-Wesley, Reading, MA, 1990, includes a long chapter on math, discussing addition, subtraction, fractions, factorials, problem solving, and "bootleg math." Holt was not a mathematician, but he was excited about math and learning. His insights are encouraging and useful for homeschooling parents. (SLS)

Math Facts: Survival Guide to Basic Mathematics, 2nd Edition, Theodore John Szymanski, PWS Publishing, Boston, 1995. A handy comprehensive reference guide that can be very useful to homeschooling parents who are rusty on the math techniques, definitions, formulas, and so on. (SLS)

Natural Math, www.naturalmath.com/index.html. This is a very unusual site dedicated to giving a new perspective on learning math and learning in general. Lots of useful and interesting discussions and helpful ways to learn multiplication tables and other things. I highly recommend it! (SLS)

Pennsylvania Homeschoolers (newsletter), R.D. 2, Box 117, Kittanning, PA 16201, www.pahomeschoolers.com. Lots of good articles and reviews of materials. There is usually a children's section called "The Backpack" with articles written by homeschoolers, and a "Math by Kids" section of math problems created by homeschooling students. The Web site has a catalog of games, textbooks, and other resources appropriate to homeschoolers and other families. (SLS)

Sassafrass Grove Math Page, www.angelfire.com/mo/sasschool/math.html. Especially for homeschoolers, a collection of math links and a list of math software for K-12. (SLS)

The Three R's at Home, Howard and Susan Richman, Pennsylvania Homeschoolers, Kittanning, PA, 1988. The math chapter is mostly elementary-school level with good descriptions of how to use Cuisenaire rods and how to relate math to everyday life. Available from PA Homeschoolers. (SLS)

Vegsource Homeschool Talk & Swap, www.vegsource.com/homeschool, is a large site that has discussion boards for many aspects of homeschooling including a "Math & Reading Board," swap boards where you can buy and sell materials, and a long (though disorganized) list of links to suppliers of materials and other homeschool-related Web sites. (SLS)

Appendix D: Catalogs and Web Sites for All Parents

Activity Resources Company, Inc., P.O. Box 4875, Hayward, CA 94540, 510-782-1300, www.activityresources.com, info@activityresources.com. A good collection of manipulatives and books. (SLS)

Art of Problem Solving, www.artofproblemsolving.com. A Web site with online enrichment courses for students on a variety of math topics and a math discussion board, also for students. These are primarily designed for high schoolers but are also appropriate for precocious younger students. Offers the *Art of Problem Solving* books (see the Problem Solving section of the annotated bibliography). (SLS)

Carolina Biological Supply Co./Carolina Math, 2700 York Rd., Burlington, NC 27215-3398, 800-334-5551, www.carolina.com/math/index.asp. Lots of excellent and sometimes hard-to-find math books, games, charts, and other items. (SLS)

Creative Publications, Wright Group/McGraw-Hill, 12600 Deerfield Parkway, Suite 425, Alpharetta, GA 30004, 800-648-2970, www.creativepublications.com. Offers excellent manipulatives and problem books. Sells Fraction Factory and Decimal Factory.

Cuisenaire, 500 Greenview Ct., Vernon Hills, IL 60061, 800-445-5985, www.etacuisenaire.com/index.htm. A source of some books and videos and many math manipulatives, especially Cuisenaire rods, which help students visualize number concepts using rods of various lengths, each length associated with a different color.

Dale Seymour (a division of Pearson Learning), 4350 Equity Drive, P.O. Box 2649, Columbus, OH 43216-2649, 800-526-9907, http://plgcatalog.pearson.com. Many math books and manipulatives. Their Circle Master compasses are especially good; they are easy to set, hold, and use. (SLS) Cuisenaire and Dale Seymour are now partners in publishing a joint catalog that includes both science and math resources for elementary and secondary levels.

Delta Education, P.O. Box 3000, Nashua, NH 03061-3000, 800-442-5444, www.deltaeducation.com. Math and science books, manipulatives, and other tools. Several pages of math games, including Muggins! and Mancala (oware).

Gnarly Math, www.gnarlymath.com, is a Web site devoted to promoting math in a friendly way. It shows ways to learn math at many levels, and it provides excellent and well-screened links to both math and non-math educational sites. (SLS)

Jacobs Publishing, 3334 East Indian School Road, Suite C, Phoenix, Arizona 85018, 800-349-1063, www.jacobspublishing.com. Unusual math books and other materials: books for use with graphing calculators, Dover books, Escher and Einstein t-shirts, posters, ties, materials for designing tesselations, puzzles, games, and workbooks. (SLS)

Marilyn Burns Newsletter, www.mathsolutions.com/newsletter. This site has many features by one of our country's leading math educators of elementary school children.

The Math Forum, http://mathforum.org, provides an answering service and many other resources.

Math Products Plus, P.O. Box 64, San Carlos, CA 95070, 650-593-2839, www.mathproductsplus.com. Books and calendars by Theoni Pappas and some other authors. Also t-shirts and posters about mathematics. My personal favorites are the mathematical playing cards! (SLS)

Mathematical Olympiads for Elementary and Middle Schools, 2154 Bellmore Ave., Bellmore, NY 11710-5645, 516-781-2400, www.moems.org, moes@I-2000.com. See the description in the "Problem Solving" section of the annotated bibliography. (SLS)

Spectrum/NASCO, 125 Mary Street, Aurora, Ontario L4G 1G3, 800-668-0600, www.spectrumed.com, speedu@attglobal.net. A great selection of math manipulatives and other supplies. (SLS)

Appendix E: Games

Chess from Europe, Go from Asia, and Oware (or Mancala) from Africa are easy enough for young children to learn to play, but they challenge the best human minds after decades of practice. Children who enjoy one (or more) of these can begin a lifetime of pleasure and may enjoy linking up with international organizations.

Family Math. See the first section of the annotated biography.

Family Pastimes, RR 4, Perth, Ontario, Canada, K7H 3C6, 613-267-4819, www.familypastimes.com, fp@superaje.com. A company that designs and sells excellent cooperative games. Although these wonderful games are not overtly mathematical, they promote teamwork and problem-solving skills, both of which are very valuable in math and elsewhere. (SLS)

Muggins! Math, 4860 Burnt Mtn. Rd., Ellijay, GA 30540, 800-962-8849, www.mugginsmath.com, muggins@mugginsmath.com. Muggins is a fun game that also teaches children their math facts and entices them into problem solving. Knock-Out and OPPs are two other excellent games from the same company. (SLS)

SET: The Family Game of Visual Perception, www.setgame.com. Finding patterns is what this game is all about. Any age can play, and frequently children can beat adults at this game. The game can also be played as solitaire or cooperatively. The Web site for SET has demos and shareware versions, so you can try it out. (SLS)

The 24-game, www.24game.com. Like Muggins, this game gives practice with basic math facts while engaging students in some out-of-the box thinking. Most of the decks have three levels of difficulty, so it is possible for people on different math levels to play together. There are decks for everything from addition and subtraction through algebra and variables. This game has caught on with children and their schools, and now regional tournaments are held all over the United States. (SLS)

Annotated Bibliography

Many books about math and math-related subjects are probably available at your local library. If you look for the ones listed here, you may find other excellent ones nearby. Enjoy browsing with your child!

Math for Children

There are many puzzle and game books in your local library—and, hopefully, in your child's school library too. Enjoy! Leave them around for your child to browse through. You can buy others from the catalogs and Web sites listed elsewhere.

Too many excellent storybooks teach math to even begin listing them. Actually, you can use almost every storybook to teach math if you approach it as *Math Power* suggests. Two of my favorites are *The Doorbell Rang* (Pat Hutchins, Greenwillow Books, New York, 1986) and *How Much Is a Million?* (David M. Schwartz, Lothrop, Lee, and Shepard Books, New York, 1985), but many others are truly excellent.

Anno's Math Games, Anno's Counting House, Anno's Counting Book, Anno's Hat Tricks, Anno's Mysterious Multiplying Jar, Socrates and the Three Little Pigs, and others by Mitsumasa Anno. These are great books for inspiring kids and adults to think more about math ideas in new and interesting ways. Even the most math-fearful child couldn't help but be drawn into these books! (SLS)

Count on Your Fingers, African Style, Claudia Zaslavsky, HarperCollins, 1980, Black Butterfly Children, 2000. Describes how finger counting is used for communication of price and quantity in an East African market place.

The Family Math books are great for parents and children enjoying math together. As described on pages 239-40, they are even better when used by groups of families with a specially prepared teacher. All are published by the Lawrence Hall of Science, University of California at Berkeley, Berkeley, CA 94720-5200, 510-642-1823, 800-897-5036, www.lawrencehallofscience.org/equals, equals@berkeley.edu.

Family Math, Jean Kerr Stenmark, Virginia Thompson, and Ruth Cossey, 1986. A great collection of games and activities for all ages.

Family Math II: Achieving Success in Mathematics, Grace Dávilla Coates and Virginia Thompson, 2003. More of the same after more years of experience.

Family Math for Young Children: Comparing, Grace D. Coates and Jean K. Stenmark, 1997. Games for Pre-K through third grade.

Family Math—The Middle School Years, Algebraic Reasoning and Number Sense, Virginia Thompson and Karen Mayfield-Ingram, 1998. Games for grades 5–8.

Get It Together, Tim Erickson, 2005. A collection of over 100 math problems for three to six people to solve together. The problems vary widely in difficulty, but each problem has six clue cards that together provide the needed information to solve the problem.

I Hate Mathematics, Little Brown and Company, 1975; *The Book of Think,* Little Brown and Company, 1976; *Math for Smarty Pants,* Little Brown and Company, 1982; *The $1.00 Word Riddle Book,* Marilyn Burns Education Associates, Sausalito, CA, 1990; and others by Marilyn Burns are delightful. You can't go wrong with a Marilyn Burns book.

Math for Adults

Innumeracy: Mathematical Illiteracy and Its Consequences, Hill and Wang, New York, 1988, and *Beyond Numeracy: Ruminations of a Numbers Man,* Vintage, New York, 1992, both by John Allen Paulos. The first was a bestseller. Both discuss a variety of math topics accessible to non-mathematicians in a very readable style.

The Mathematical Experience, Philip Davis and Reuben Hersh, Houghton Mifflin, Boston, 1981. Traditional mathematical material written in an entertaining, enlightening way for the general adult reader.

Mathematics: A Human Endeavor: A Book for Those Who Think They Don't Like the Subject, Harold R. Jacobs, W. H. Freeman and Company, San Francisco, 1970, 1994, 2005. Includes probability, large numbers, scientific notation, and topology. A user-friendly text for students age 12 and up and for adults. Instructors' manual is also available.

Number: The Language of Science, Tobias Dantzig, Macmillan, New York, 1930, reprinted by Pi Press, New York, 2005. This was the first widely selling book of this type. Generations of readers have enjoyed its lucid explanations of some of the subtleties in mathematics.

The Heart of Mathematics: An Invitation to Effective Thinking, Second Edition, Edward B. Burger and Michael Starbird, Key College Publishing, Emeryville, CA, 2005. This text is written in an entertaining, often funny style that charmed my college non-majors and might please other adults. Its voluminous pages include the standard non-majors' topics and recent additions such as fractals, infinity, voting theory, and decision-making.

Powers of Ten, Philip Morrison and Phylis Morrison, Scientific American Library, distributed by W. H. Freeman and Company, New York, 1982. Based on a 20-minute film by the same name by The Office of Charles and Ray Eames, this is a picture book illustrating the effect of increasing and decreasing magnification by ten, starting with a man asleep on a blanket. Multiplying repeatedly by ten, we eventually see groups of galaxies, and dividing, we see images of electrons. Effectively helps children and adults understand concrete applications of abstract ideas.

Mathematics for Human Survival, Patricia Clark Kenschaft, Whittier Press, Island Park, NY, 2002. Written as a text for college non-majors, its exercises and examples use all real numbers from health, peace, and environmental issues. It includes a list of all the countries in the world with their vital statistics, information that is used as a basis for many of the book's problems. This could be used by inquiring minds in middle school or younger.

The Only Math Book You'll Ever Need, Stanley Kogelman and Barbara Heller, Dell, New York, 1988. Despite its arrogant title, this is a good reference about math that's useful for taxes, banking, loans, investments, tips, discount, cars, international exchange rates, the consumer price index, home and cooking measurement, and other topics needed by citizens. It serves "the math phobic" well, to quote the authors' phrase.

The Joy of Mathematics, World Wide Publishing/Tetra, San Carlos, CA, 1989, and *More Joy of Mathematics*, World Wide Publishing/Tetra, San Carlos, CA, 1991, both by Theoni Pappas. These explore hundreds of topics that are related to math in some way. Some topics require algebra or higher math, but many do not. Both books have lots of illustrations and complete indexes. Theoni Pappas has authored many other books about math and also produces a mathematical calendar. Two of her newer books are *Mathematical Footprints: Discovering Mathematical Impressions All Around Us*, World Wide Publishing/Tetra, San Carlos, CA, 2000, and *Math Stuff*, World Wide Publishing/Tetra, San Carlos, CA, 2002. These books are fun to read and browse. Some chapters are good for children. (SLS)

Statistics: Concepts and Controversies, David S. Moore, W. H. Freeman and Company, New York, 1970, 1985. This is an entertaining book with many illustrations about the statistics you need to read magazines intelligently.

How to Lie With Statistics, Darrell Huff, Norton, New York, 1954. A tiny, funny, often-reprinted classic with a self-explanatory title.

Learning and Teaching Math

Knowing and Teaching Elementary Mathematics: Teaching Understanding of Fundamental Mathematics in China and the United States by Liping Ma, Lawrence Erlbaum Publishers, Mahwah, NJ, 1999. May be the most influential book about the teaching of math to appear during the past decade. Written after extensive studies in both China and the United States by someone totally bilingual, it reveals startling differences in the preparation and nurturing of teachers in the two countries, along with inspiring ways that Chinese children (and teachers) are taught. It has been well received by both sides of the "Math Wars."

Radical Equations: Mathematics and Civil Rights by Robert Moses and Charles E. Cobb, Jr., Beacon Press, Boston, 2001. Has not received as much attention, but it tells what can be done to help under-prepared middle school students be ready for high school and college mathematics. Moses was an advanced mathematics student and teacher whose career was twice deflected by civil rights. He now leads The Algebra Project, which reaches 10,000 African American middle schoolers a year. The book is a dramatic story of personal and social change, but it also outlines Moses' successful method for stimulating mathematical growth.

The Learning Gap: Why Our Schools Are Failing and What We Can Learn from Japanese and Chinese Education, Harold Stevenson and James Stigler, Simon and Schuster, New York, 1992. Compares the mathematical achievement of three cities in the United States, Japan, and Taiwan with appalling conclusions for an American reader. This book was one of the reasons that *Math Power* was written, and it is quoted frequently.

The Teaching Gap: Best Ideas from the World's Teachers for Improving Education in the Classroom, James Stigler and James Hiebert, Free Press, New York, 1999. Compares how teachers are prepared and nourished in Japan, Taiwan, and the United States, under the assumption that the culture of teachers affects their students' achievement. They describe in detail the custom of "lesson study," whereby a group of teachers prepares a lesson together, one of them teaches it to his or her class with the others watching, they revise the lesson together using what they learned during that lesson, and another teacher in the group teaches the lesson with the others watching. This process can continue until the lesson is well "polished."

Educating Hearts and Minds: Reflections on Japanese Preschool and Elementary Education, Catherine Lewis, Cambridge University Press, New York, 1995. Written by a scholar of Japanese education for 14 years. Totally fluent in Japanese, the author bases her writing on many interviews and observations of Japanese classrooms and those involved in Japanese education. The work focuses on preschool and early-elementary education, where there are classes of 30 to 40 children divided into diverse cooperative groups. She concludes with questions about what Americans might learn from Japanese practices.

Mathematics Their Way, Mary Baratta-Lorton, Center for Innovation in Education, Saratoga, CA, 1976. One of the most popular teaching resources for primary grade teachers. It includes over 200 activities that teach basic mathematical understanding. Now available in a 20th anniversary edition from the Center for Innovation in Education, Inc., Saratoga, CA, 800-395-6088, www.center.edu.

The Challenge to Care in Schools: An Alternative Approach to Education, Second Edition, Nel Noddings, Teachers College Press, New York, 2005. By a former high school mathematics teacher and mother of ten (five biological), who later became a professor of education at Stanford University and then president of the National Academy of Education. She has written many other books and papers about education. In this one she lists many types of caring and explains why each is important.

Mathsemantics: Making Numbers Talk Sense, Edward MacNeal, Penguin (non-classics) reprint edition, New York, 1995. A well-written, insightful, and fascinating book that challenges our way of thinking about math and gives some concrete guidelines for learning how to think in a mathematical way. (SLS)

Young Children Reinvent Arithmetic, Implications of Piaget's Theory, Constance Kamii and Leslie Baker Housman, Teachers College Press, New York, 1985. Kamii has written many books for educators that may interest parents. This is the best known.

Speaking Mathematically: Communication in Mathematics Classrooms, David Pimm, Routledge, New York, 1987, 1989. Despite its subtitle, this little book provides insight for anyone attempting to communicate mathematics. Some examples are too abstruse for most parents, but many provide a next step beyond *Math Power's* Chapter 8. Good discussion of images and metaphors.

Beyond Facts & Flashcards: Exploring Math with Your Kids, Jan Mokros, Heinemann, Portsmouth, NH, 1996. Gives parents concrete examples of ways to talk math with their children and encourage their natural interest in numbers, geometry, patterns, and other math topics. Primarily for use with children age 12 and under. (SLS)

Experiencing School Mathematics: Teaching Styles, Sex and Setting, Jo Boaler, Open University Press, Buckingham and Philadelphia, 1997. Reports on a survey of two schools in England with extremely different teaching philosophies with similar students at the beginning. The school that gave students three weeks to do open-ended problems scored significantly higher in the national exams than the one that followed the traditional "sage on the stage" approach, segregated by levels of middle school achievement. The girls in the top level were the ones who seemed to suffer most by experiencing the traditional instead of open-ended approach.

The following books read like novels and tell how two recent U.S. teachers empower their students to learn spectacularly compared to surrounding students:

Marva Collins' Way, Marva Collins and Civia Tamarkin, G.T. Putnam's Sons, New York, 1982, 1990.

Escalante: The Best Teacher in America, Jay Mathews, Henry Holt and Company, New York, 1988.

Problem Solving

How to Solve It, George Polya, Princeton University Press, Princeton, NJ, 1945, 1957, 1971. This classic has been translated into dozens of languages and has sold over a million copies. It indicates how to excel at high school mathematics and speaks to adults whose math knowledge is at that level or above. Polya was a great research mathematician whose biography is included in volume 2 of the child's series *Mathematicians Are People, Too* listed later.

The Art of Problem Solving, Volume 1: The Basics and *Volume 2: And Beyond*, Sandor Lehoczky and Richard Rusczyk, Greater Testing Concepts, 1993. The authors address most of the basic topics in Algebra I and II and Geometry in volume 1, but in a completely different way from traditional texts. They use problems from various contests—Mathcounts, AHSME, USAMTS, the Mandelbrot Competition (their own contest), and others—to illustrate the concepts they present. This book would not be a good substitute for the traditional textbooks, but it is an excellent supplement because of its extensive use of applications. The critical and creative thinking processes that are encouraged in this book are valuable for every student. A manual with complete solutions for all the problems is available and recommended. Although written for high school students and adults, this book appeals to children who are really advanced, as some homeschooled children are by late elementary school. Available from www.artofproblemsolving.com. (SLS)

Creative Problem Solving in School Mathematics, Math Olympiads, Bellmore, NY, 2005, and *Mathematical Olympiad Contest Problems for Children,* Glenwood Publications, East Meadow, NY, 1997, both by George Lenchner. These books are based on contests that schools can run in-house for children in grades 4–7. The contests are available nationally but do not feed into higher-level competition than the local school, so they can be kept low-key. Both books provide complete solutions. The former has suggestions for teaching problem solving and gives strategies with many examples. Some problems are organized by topics. It includes 20 complete Olympiads with five problems each. The second book includes these problems and 60 more complete Olympiads, all with answers, hints, and complete solutions. (SLS)

Susan Schaeffer reflects my sentiments when she writes, "If the contest is handled well, there is usually a group discussion of each of the five problems that the students work on in a round (one round a month), so the participants have the opportunity to get better at talking about math, too ... the awards aspect is kept low-key in most circumstances, I think. Because of this emphasis on process (as opposed to results), there is little outside pressure on students, which in my opinion is about the only way I could approve of a contest for this age group. I have misgivings about math contests in general, but when students are not pressured to perform, I think it can be a great way to learn math problem solving, especially of material outside the normal curriculum. It also provides a way for students who have a strong interest in math ... to meet and form a community with other students like themselves—an important consideration, as students who enjoy math are often teased and derided in a culture where being bad at math is quite acceptable and often considered normal." I would add that this community may be especially important during the insecure early adolescent years.

Math by Kids, edited by Susan Richman, Pennsylvania Homeschoolers, Kittanning, PA, 1996. A compilation of word problems written by children and teens in eleven different categories with varying levels of difficulty from easy to mindbender. Complete solutions are included, as well as tips for teaching problem solving. Available from Pennsylvania Homeschoolers, listed earlier in Appendix C. (SLS)

Mathematical Investigations, Dale Seymour Publications, Palo Alto, CA, 1990. This is a three-volume set of "situational lessons" on topics such as patterns, photography, sports math, using maps, and understanding finance, among others. Designed for students grades 7–12, but perhaps appropriate for some younger ones. (SLS)

Problem Solving Strategies: Crossing the River with Dogs and Other Mathematical Adventures, Second Edition, Ken Johnson and Ted Herr, Key Curriculum Press, Emeryville,

CA, 2001. This is chock full of interesting patterns and would be an excellent supplement to any curriculum. It is a user-friendly book; there are lots of detailed examples with hand-drawn diagrams, charts, and tables. Because of the wide range of problems included, it is appropriate for students from early middle school through high school. (SLS)

Psychology

Fear of Math: How to Get Over It and Get On with Your Life!, Claudia Zaslavsky, Rutgers University Press, New Brunswick, NJ, 1994. Includes many stories about the math-anxious and the math-confident, especially those who moved from the first group to the second.

Dyscalculia, www.dyscalculia.org. About math anxiety and other problems that prevent people from learning math well. It includes a comprehensive list of references. (SLS)

Overcoming Math Anxiety, Sheila Tobias, Norton, New York, 1978, 1993. A classic on how to return as an adult to the world of the mathematically functional.

Why Bright Kids Get Poor Grades: And What You Can Do About It, Sylvia Rimm, Crown Publishers, New York, 1995. Explains in detail how parents, teachers, and kids working together can rescue youngsters from remaining underachievers.

Equity

Teaching Tolerance, 400 Washington Avenue, Montgomery, AL 36104, 334-956-8200, mailed twice a year at no cost to educators. This magazine is full of ideas for making classrooms and society more tolerant of all varieties of people. It also offers many resources, including the book *Words Are Not for Hurting* for toddlers and preschoolers and the video with *Let's Get Real* for middle schoolers about bullying and name-calling with interviews of both victims and bullies, and a discussion guide.

Africa Counts: Number and Pattern in African Culture, Claudia Zaslavsky, 1973, reprinted steadily and now in paperback, Lawrence Hill Books, Chicago, 1999. This classic opened Western eyes to the heritage of African mathematics. It is based on trips by Zaslavsky and her son and friends. Since then, she has had 12 other books published (most still in print) based on international mathematics.

Math Games & Activities from Around the World, Claudia Zaslavsky, Chicago Review Press, 1998. Also from the same author and press: *More Math Games & Activities from Around the World*, 2003; *Number Sense and Nonsense: Building Math, Creativity, and Confidence through Number Play*, 2001.

Multicultural Mathematics: Teaching Mathematics from a Global Perspective, David Nelson, George Gheverghese Joseph, and Julian Williams, Oxford University Press, New York, 1993. Explores ways of helping school children understand the universality of mathematics.

The Multicultural Math Classroom: Bringing in the World, Claudia Zaslavsky, Heineman, Portsmouth, NH, 1995. Inspires cooperation, creativity, and critical thinking.

Savage Inequalities: Children in America's Schools, Jonathan Kozol, Crown Publishers, New York, 1991. A passionate exposé comparing nearby rich and poor school districts. Kozol is one of the most articulate advocates for uniform funding of schools nationwide.

Other People's Children: Cultural Conflict in the Classroom, Lisa Delpit, The New Press, New York, 1995. An analysis of the subtle ways that well-meaning teachers may unintentionally undermine their minority students' education.

Failing at Fairness: How America's Schools Cheat Girls, Myra and David Sadker, Scribner, New York, 1994. A chilling exposé with a long and touching chapter on how we are harming boys, too.

How Schools Shortchange Girls: The AAUW Report, Marlowe and Company, New York, 1992. A widely publicized, well-researched report that drew attention to a widespread problem with considerable attention to the unequal opportunities for girls to learn mathematics in U.S. schools.

Standardized Testing

The Fractured Marketplace for Standardized Testing, Walter Haney, George Madaus, and Robert Lyons, Kluwer Academic Publishers, Boston, 1993. The classic book questioning the costs (visible and invisible), reasons, and results of testing.

One Size Fits Few: The Folly of Educational Standards, Susan Ohanian, Heinemann, Portsmouth, NH, 1999. A passionate plea for treating children as individuals. She is a former elementary school teacher, and the book is full of revealing anecdotes.

Contradictions of School Reform: Educational Costs of Standardized Testing, Linda McNeil, Routledge, New York, 2000. Written by a professor of education at Rice University who has spent much time in the Houston, Texas, public schools, whose presumed success has become the model for national policy. She sheds doubt on that "success" while raising other very important questions.

The Case Against Standardized Testing: Raising the Scores, Ruining the Schools, Alfie Kohn, Heinemann, Portsmouth, NH, 2000. Recounts the damage standardized testing does to education, especially to already disadvantaged groups, explores questions about the validity of scores, emphasizes that standardized tests "are not like the weather, something to which we must resign ourselves," and demonstrates how parents, teachers, and students can resist the impact of testing and create classrooms that focus on learning.

FairTest: National Center for Fair and Open Testing, www.fairtest.org, 342 Broadway, Cambridge, MA 02139, 617-864-4810. A national watchdog organization eager to improve U.S. testing practices. Provides a newsletter, bibliography, and provocative original publications.

Biography and History

Mathematicians Are People, Too, Luetta Reimer and Wilbert Reimer, Dale Seymour Publications, Palo Alto, CA, 1995. Charmingly illustrated, beguiling biographies for children of eleven male and four female mathematicians, including George Polya (1887-1985), mentioned earlier in the bibliography about teaching and learning mathematics.

Math Equals, Teri Perl, Addison-Wesley, Menlo Park, CA, 1978. Biographies and pictures of historical women in mathematics and activities for children illustrating their research. Geared toward late elementary school.

Women and Numbers: Lives of Women Mathematicians, Teri Perl, Wide World Publishing/Tetra, San Carlos, CA, 1993. Includes activities in every chapter.

There are several books about women mathematicians written for adults.

Women in Mathematics, Lynn M. Osen, MIT Press, Cambridge, MA, 1974.

Women of Mathematics: A Bibliographic Sourcebook, edited by Louise Grinstein and Paul Campbell, Greenwood Press, Westport, CT, 1987. Short biographies and publication summaries of many women mathematicians, living and dead.

She Does Math! Real-Life Problems from Women on the Job, edited by Marla Parker, Mathematical Association of America, Washington, DC, 1995. Short biographies of women in a wide variety of fields and accounts of how they use mathematics in their careers.

Women in Mathematics: The Addition of Difference, Claudia Henrion, Indiana University Press, Bloomington and Indianapolis, 1997. Nine in-depth biographies based on extensive interviews amid essays about myths about mathematics and women in mathematics.

Notable Women in Mathematics: A Biographical Dictionary, edited by Charlene Morrow and Teri Perl, Greenwood Press, Westport, CT, 1998. Short biographies of many women mathematicians, living and dead.

Women Becoming Mathematicians: Creating a Professional Identity in Post-World War II America, Margaret A. M. Murray, MIT Press, Cambridge, MA, 2000. This is a narrative integrating interviews of survivors and documents about the women in the United States who earned a Ph.D. in mathematics between 1940 and 1960.

Change is Possible: Stories of Women and Minorities in Mathematics, Patricia Clark Kenschaft, American Mathematical Society, Providence, RI, 2005. Includes stories of dozens of living mathematicians based on interviews and some historical stories.

Notes

Preface to the Revised Edition

1. "A New Look at Public and Private Schools: Student Background and Mathematics Achievement," *Phi Delta Kappa*, May 2005, Survey from the National Briefing Service supported by the National Security Agency and Exxon Mobil Foundation, as quoted in the *Christian Science Monitor*, May 10, 2005, p. 11.

2. "1.1 Million Homeschooled Students in the United States in 2003," National Center for Educational Statistics, U.S. Department of Education Institute of Educational Sciences, July 2004, NCES 2004-115.

3. "One Third of a Nation: Rising Dropout Rates and Declining Opportunities," Educational Testing Service, Princeton, NJ, February 2005. www.ets.org/rsearch/pic, p. 7.

4. Ibid., p. 7.

5. Ibid., p. 8.

6. Ibid., p. 11.

7. Ibid., p. 13.

8. Robert Balfanz and Nettie Legters, "Locating the Dropout Crisis," Center for Social Organization of Schools, Johns Hopkins University, June 2004, first page of the Executive Summary.

9. Confirmed by Catherine Lewis in a personal email on June 24, 2005.

10. McCollum, P., and A. Cortez, O. H. Maroney, F. Montes, "Failing Our Children—Finding Alternatives to In-Grade Retention," Intercultural Development Research Association, San Antonio, 1999. Quoted in "Education Statistics," p. 3, www.idra.org/Research/edstats.htm#ingrade.

11. Intercultural Development Research Association, "Statistics and Data on Dropout Prevention," www.idra.org/Newslttr/Fieldtrp/2000/Statsoct.htm, p. 3. Quoting Roy L. Johnson, "Missing: Texas Youth—Cost of School Dropouts Escalates," IDRA Newsletter, October 2001.

12. Ibid., October 2001.

13. Stephen W. Raudenbush, "Schooling, Statistics, and Poverty: Can We Measure School Improvement?," ETS, Princeton, 2004.

14. www.fairtest.org/arn/washington%20press%20release.html.

Chapter 1

1. Mathematical Sciences Education Board, *Everybody Counts, A Report to the Nation on the Future of Mathematics Education* (Washington, D.C.: National Academy Press, 1989), 77.

2. P. C. Kenschaft, "Black Women in Mathematics in the United States," *American Mathematical Monthly*, 88:8 (October 1981): 600. Reprinted with photos and post-script in the *Journal of African Civilization*, 4:1 (April 1992): 76.

3. The respondents included 17 in higher education, 23 in precollege education, four retired, two full-time graduate students, and 29 others in a wide variety of careers, including high-level computer analysis and sales, financial and budget analysis, technical management, and "in rotation" positions for top management. P. C. Kenschaft, "What Are They Doing Now? Careers of 75 Black Mathematicians," *UME Trends* (May 1990): 4.

4. P. C. Kenschaft, "Successful Black Mathematicians of New Jersey," *UME Trends* (December 1989): 4.

5. P. C. Kenschaft, "Black Women in Mathematics," *Newsletter of the Association for Women in Mathematics* (September- October 1988): 5-7.

6. Council of Chief State School Officers, *State Education Policies on K-12 Curriculum, Student Assessment, and Teacher Certification: 1995, Results of a 50-State Survey* (Washington, D.C.: Council of Chief State School Officers, 1995): 15.

7. Earlier this century, people even taught dogs and horses to do math. Gesturing with their paws and hoofs, these animals were presumed to be able to count and do sums. Some professors verified the claims of devoted owners. But claims of horses taking square roots of large numbers strains credibility. In *The Magic of Numbers*, author Robert Tocquet (Hal Leighton Printing Co., Beverly Hills, Calif., 1957, 1976 edition) concludes that such mathematical feats resulted from hidden, and perhaps involuntary, communications from people to these animal marvels.

   Tocquet does concede that chickens can be trained to recognize the number 3 but not 4 and that parrots and crows can recognize numbers up to 7. What if similar time were invested in all human children? (See also comments about number sense in non-humans in Chapter 11.)

8. Harold W. Stevenson and James W. Stigler, *The Learning Gap: Why Our Schools Are Failing and What We Can Learn from Japanese and Chinese Education* (New York: Simon & Schuster, 1992), 220.

   This readable book, quoted throughout *Math Power*, is a fascinating analysis of Japanese, Chinese, and U.S. education. It details the authors' extensive international studies comparing schools in Minneapolis, Chicago, Sedai in Japan, and Tiapai and Beijing in the People's Republic of China. Their investigations included comparisons of student achievement tests; interviews with students, teachers, and parents; classroom observations; and open-ended surveys. The studies included thousands of subjects.

9. Ibid., 216.

10. Michael Fellows and Neal Koblitz, "Combinatorially Based Cryptography for Children (and Adults)," *Proceedings of the SE Conference on Graphs, Combinatorics, and Computing* (1993).

11. For example, *Money* magazine's *Forecast 1997* issue says, "Again, to no surprise, the fastest-growing occupation in '97 will be the same one that's topped *Money's* annual hot-job list for three years running: technically trained pros, such as computer programmers, software designers, and engineers" (p. 123). Year after year, mathematics and math-intensive careers lead such lists.

12. John Allen Paulos, *Innumeracy* (New York: Hill and Wang, 1988) 10. This charming (and short) bestseller is a good follow-up to *Math Power*.

13. This paragraph was written before the publication of *Splintered Vision*, a report on an international study that corroborates my assertions here. See Appendix A.

14. Michael Fellows and Neal Koblitz, "Combinatorially Based Cryptography for Children (and Adults)," *Proceedings of the SE Conference on Graphs, Combinatorics, and Computing* (1993): 16-17. This paper describes specific games and activities in modern computer science that children can enjoy *without* computers.

Chapter 2

1. President Bill Clinton, at the National Education Summit, Palisades, New York, March 27, 1996, as quoted in "President Urges Standards That Count," the *American Educator* (Spring 1996): 8-12.

2. Jay Mathews, *Escalante: The Best Teacher in America* (New York: Henry Holt and Co., 1988), 301.

3. Mathews, *Escalante:* This book reads like a novel, telling the story of a determined man coping with a frustrating system. Although many math educators expressed concern over some of the methods Escalante used, his achievements are undeniable, and they prove that *ganas* can overcome a broken home and a discouraging work environment. One of Escalante's colleagues had comparable success with a somewhat modified version of Escalante's methods after Escalante defied the odds and succeeded.

4. Gina Kolata, "At Last, Shout of 'Eureka!' In Age-Old Math Mystery," *New York Times*, 24 June 1993, A1, D22; Gina Kolata "Math Whiz Who Battled 350-Year-Old-Problem," *New York Times*, 29 June 1993, C1, C11; Kim A. McDonald, "Princeton Professor Appears to Have Proved Fermat's Last Theorem," *The Chronicle of Higher Education,* 7 July 1993, A8, A17; James Gleick, "Fermat's Theorem," *New York Times Magazine*, 3 October 1993, 52-53; Kitta MacPherson, "Waiting for Fermat: Princeton Math Prof Must Prove His Proof," *Sunday Newark Star-Ledger*, 7 November 1993, sec. 1, 47; Anthony F. Shannon, "Proof Not Positive for Princetonian Hailed for Solving Famous Theorem," *Star-Ledger*, 11 December 1993, 11; Gina Kolata, "How a Gap in the Fermat Proof Was Bridged," *New York Times*, 31 January 1995, B5, B8.

5. *Merriam-Webster Pocket Dictionary* (New York: Simon & Schuster, 1974), 39.

6. Kenneth Appel and Wolfgang Haken, "The Four-Color Problem," in *Mathematics Today*, ed. Lynn Steen (New York: Random House, 1978), 153-80.

7. Michael Pollan, *Second Nature* (New York: Dell Publishing, 1991), 145.

8. Elizabeth Chambers Patterson, "Mary Fairfax Greig Sommerville (1780-1872)," 1987 in *Women of Mathematics, A Bibliographic Sourcebook*, ed. Louise S. Grinstein and Paul J. Campbell (New York: Greenwood Press), 208-16.

9. Mary W. Gray, "Sophie Germain (1776-1831)," in *Women of Mathematics, A Bibliographic Sourcebook*, ed. Louise S. Grinstein and Paul J. Campbell (New York: Greenwood Press), 208-16.

Chapter 3

1. Beatrice Stillman, *Sofya Kovalevskaya: A Russian Childhood* (New York: Springer-Verlag, 1978), as quoted on 35 (VIP, 314).

2. Hugo Rossi, "More on Researchers and Education," *Notices of the American Mathematical Society*, 37:6 (July/August 1990): 6-7.

3. Charles Sykes, *Dumbing Down Our Kids: Why American Children Feel Good About Themselves but Can't Read, Write, or Add* (New York: St. Martin's Press, 1995), 58.

   Sykes distinguishes between "self-esteem," which he claims is taught directly to children through statements that they are important, and "confidence," which is gained through achievement. I perceive a large "gray area" in between—perhaps an entire continuum between self-esteem and confidence—but I think Sykes' distinction is interesting. I don't share his antipathy toward self-esteem, but I do believe confidence is much more valuable for learning mathematics. Confidence empowers us to take risks.

4. See the quote in the fifth paragraph of Chapter 12.

5. Susan Ohanian, *Garbage Pizza, Patchwork Quilt, and Math Magic* (New York: W. H. Freeman and Co., 1992). This is a truly inspiring book, full of specific, imaginative ideas.

6. Marva Collins and Civia Tamarkin, *Marva Collins' Way* (New York: G. P. Putnam's Sons, 1982, 1990).

7. Alexis Jetter, "Mississippi Learning," *New York Times Magazine*, 21 February 1993, 28-35, 50-51, 64, and 72. Cover article with cover title, "We Shall Overcome, This Time with Algebra."

8. Mathews, *Escalante*.

9. For example, E. D. Hirsch, *What Your 1st Grader Needs to Know* (New York: Dell Publishing Group, Inc., 1991), 158.

10. Mary Leonhardt, *Parents Who Love Reading, Kids Who Don't* (New York: Crown Publishers, 1993), 26.

11. *Developmentally Appropriate Practices in Early Childhood Programs Serving Children from Birth Through Age 8*. Ed. Sue Bredekamp (National Association for the Education of Young Children: Washington, D.C.: 1996), 7.

12. Eleanor Duckworth, "The Having of Wonderful Ideas," *Harvard Educational Review*, 42:2 (May 1972), 115.

13. Former mathematics teacher educator Ned Noddings writes provocatively about the role of caring in education. See, for example, her chapter in *Stories Lives Tell*. Eds. Carol Witherell and Nel Noddings (New York: Teachers College Press, 1991).

Chapter 4

1. David Pimm, *Speaking Mathematically* (New York: Routledge, 1987, 1989), 99 (first sentence), 108-10.

2. E. D. Hirsch and company want children at this stage to distinguish between cardinal and ordinal numbers. That seems overly pedantic to me. If we need such distinctions (which I doubt), they more rightly belong to English lessons than math. (E. D. Hirsch holds a doctorate in English.) To me, it seems that saying, "Ed is number three in his class" is just as satisfactory as saying, "Ed is third in his class." Ordinals can use the names of cardinals, and the difference is subtle.

Chapter 5

1. Robert Davis, *Discovery in Mathematics* (Reading, Mass. Addison-Wesley, 1964), 21.

2. David Elkind, *Miseducation* (New York: Knopf, 1985), 21. "MacArthur fellows" are people who are chosen to be supported in whatever they want to do for five years by the MacArthur Foundation.

3. The writing and teaching of Glenn Doman urges parents to preserve their preschoolers' "rage to learn," which he claims is universal in young children but too often destroyed by schools. His *How to Teach Your Baby Math* (Garden City Park, N.Y.: Avery Publishing Group, 1994), which presumably has over 5 million copies in print, promotes a very rigid approach that bothers me. I wonder how much his program's apparent success is due to his specific advice and how much to other factors, including the inspiration and affirmation he gives parents in his seminars.

4. Leonhardt, *Parents Who Love Reading, Kids Who Don't*, 117.

5. Mathematician Dr. Henry Pollak remembers a middle school teacher telling him that zero was not a number. "Why not?" he asked. "It's a dividing line between the positive and negative numbers," the teacher answered.

6. Stevenson and Stigler, *The Learning Gap*, 195.

7. Ibid., 197.

8. Ibid., 188.

9. Estimation is one of the thirteen "standards" contained in the NCTM *Standards* document. We will explore it further in Chapter 11. (See also Appendix A for descriptions of the NCTM *Standards*.)

10. Neil Harvey, *Kids Who Start Ahead Stay Ahead: What Actually Happens When Your Home-Taught Early Learner Goes to School* (Garden City Park, N.Y.: Avery Publishing Group, 1994), pages 69-70. Author Neil Harvey calls this quote an "astonishing observation." His book reports (both anecdotally and statistically) the results of surveys of graduates of Glenn Doman's programs for parental teaching of very young children.

11. Paul E. Barton and Richard J. Coley, *America's Smallest School: The Family* (Princeton, N.J.: ETS Policy Information Center, 1992), 2.

12. Ibid., 23, 42.

13. Stevenson and Stigler, *The Learning Gap*, 58-59.

Chapter 6

1. Constance Kamii, *Young Children Reinvent Arithmetic* (New York: Teachers College Press, 1985), 35.

2. David Elkind, *Miseducation* (New York: Knopf, 1985), 22.

3. Glenn Doman, *How to Teach Your Baby Math* (Garden City Park, N.Y.: Avery Publishing Group, 1994), 90.

4. Harvey, *Kids Who Start Ahead and Stay Ahead*, 55-58.

5. Kenschaft, "Successful Black Mathematicians of New Jersey," 4.

6. Leonhardt, *Parents Who Love Reading, Kids Who Don't*, 115-16.

Chapter 7

1. *Everybody Counts*, 43.

2. The Hand Game is described in the excellent book *Mathematics Their Way* by Mary Baratta Lorton. (See the bibliography section, "Math for Your Children.") She recommends using Unifix Blocks, one of the popular "mathematics manipulatives" that help children learn mathematics by playing. I recommend *Mathematics Their Way* to all primary-grade teachers, including home-schooling parents. If you can afford to buy a copy, you will find many excellent activities for preschoolers. If not, get it from your library; it is surely available through interlibrary loan, and your library may buy it at your request.

3. Elkind, *Miseducation*, 165.

4. Aggie Azzolino, *Math Games for Adult and Child* (Keyport, N.J.: Mathematical Concepts, Inc., 1993). Ms. Azzolino has written several books of interest to "mathphiles."

Chapter 8

1. Eleanor Wilson Orr, *Twice As Less: Black English and the Performance of Black Students in Mathematics and Science* (New York, W. W. Norton & Company, 1987).

2. For years Miss Mason's School in Princeton taught addition and subtraction of signed numbers to kindergartners by having a large number line outdoors from –10 to +20 over which the children could step and jump. To add 3 + 2 (or 3 + (-2)), they would stand on 3 and jump two places forward (or backward). Subtraction meant "turn around first, and then jump." Indoors, there were similar but smaller number lines over which toy animals could move.

   These methods have also been used successfully in urban Miami, Florida. Miss Mason's materials are now available from the Newgrange School, 52 Lafayette Avenue, Trenton, N.J., 08610. Phone: 609-394-2255, Fax: 609-394-9467.

3. In the early days of computer translation, the saying, "The spirit is willing, but the flesh is weak," was put into the computer. After being translated into Russian and then translated back, it became, "The vodka is good, but the meat is rotten." Those of us who heard it laughed nervously, wondering if nuclear annihilation could result from such imperfect understandings among countries.

4. Intended answer: They all contain six x's.

5. If you want to read a profound exposition of the meaning of zero, and two, read *Mathematical Logic* by Bertrand Russell. One of our outstanding twentieth-century mathematician-philosophers, Russell wrote this book while in a prison as a conscientious objector during World War I. It explains to nonmathematicians answers to seemingly simple but actually extremely difficult questions. Just attempting to read this book will help you develop empathy with your first grader. Basic math is profoundly difficult; it isn't the kids' fault or the teacher's.

6. Some people wish this list included definitions, but, as noted elsewhere, definitions in basic mathematics are treacherous. Most of these words must be learned by context. If you don't know them yet, you might want to read math books, join a Family Math program (see pp. 239-40), or take a course. Children who don't learn them by the end of elementary school will have unnecessary trouble with math in high school.

7. Ibid.

8. David Pimm's excellent little book *Speaking Mathematically* (New York: Routledge Press, 1987, 1989) discusses this particular problem and its consequences on page 79. Many of the other words mentioned in this paragraph are from his chapter about special linguistic quirks of those who "speak math." Much of this book, although subtitled *Communication in Mathematics Classrooms*, would interest parents who want to pursue the ideas of mathematical language. Some of the math in *Speaking Mathematically* is far beyond that addressed in *Math Power*, but presumably, a motivated parent could simply skip those passages.

Chapter 9

1. Mathews, *Escalante*, 111.

2. See the section on "Psychology" in the bibliography for two of the best. *Math Anxiety* has become a classic, and *Fear of Math* is an excellent modern book.

3. "Parents and Mathematical Games" by Owen Tregaskis (*The Arithmetic Teacher*, March 1991, 38:7, 14-16) relates successes when a classroom teacher provided games for the parents of slower students to play with their child. Both parents and children learned math and soon invented their own math games. "Before the project, parents were concerned about the children's *ability*...after the project, they were interested in the children's *understanding, confidence*, and *enjoyment.*"

4. Dr. Goldberg, of the Education Development Center in Newton, Mass., also has noticed two other common beliefs that interfere with mathematics learning in this culture: (1) Thinking is a sign of stupidity. If you can give a quick answer, you are smart; if you reflect before answering, you are not so smart (see also Chapters 16 and 17); (2) Curiosity is a sign of stupidity. You will appear blasé if you are knowledgeable, and knowledge indicates intelligence.

5. Mathematician Dr. Henry Pollak catches supermarket machines making mistakes with alarming regularity. He does this by adding the whole dollar amounts of the prices on the items and then adding half the number of items (thereby approximating the cents part to be $0.50). If the machine is too far from his estimate, he checks the tape carefully.

   My husband, Dr. Frederick Chichester, also finds mistakes by supermarket machines, although he estimates the total price of his order by rounding up or down to the nearest dollar. He also notices mistakes in the unit pricing with distressing frequency.

   Math comfort can help one elude "sharks" in many contexts.

6. Stevenson and Stigler, *The Learning Gap*, 187.

7. Dorothy Cantor and Toni Bernay, *Women in Power: The Secrets of Leadership* (Boston: Houghton Mifflin, 1992).

8. *A History of Mathematics in America Before 1900* by David Eugene Smith and Jekuthiel Ginsburg (Chicago: Mathematical Association of America and the Open Court Publishing Company, 1934) may be the best source. *A Catalogue of All the Books Printed in the United States*, published in Boston in 1804, includes 29 "mathematical books," of which six were about astronomy and the others about arithmetic (pages 73-74). In 1803, Harvard *began* requiring "the mere rudiments of arithmetic" for admission. "Not until 1837 was arithmetic dropped from the freshmen course" (page 70). In 1823, Middlebury College, Vermont, reported there were 18 math books in its library (page 73). Indeed, until 1850, many "mathematics" professors were primarily prepared in theology (page 19).

Chapter 10

1. Stevenson and Stigler, *The Learning Gap*, 18.

2. Rudy Rucker, *Mind Tools: The Five Levels of Mathematical Reality* (Boston: Houghton Mifflin Co., 1987), 55.

3. Tobias Dantzig, *Number, the Language of Science* (New York: Macmillan, 1930, 1954), 1-3.

4. Ibid., 4-5.

5. E. D. Hirsch, *What Your 1st Grader Needs to Know*, 170.

6. The geometric interpretations of the Associative and Distributive laws appear at the end of Chapter 12.

7. An excellent exposition of multiplication in its many guises, including area, is Marilyn Burns' "replacement unit" *Math by All Means: Multiplication, Grade 3*, available from Cuisenaire (see Appendix C or call 800-237-0338).

8. E. D. Hirsch, ed., *What Your 2nd Grader Needs to Know* (New York: Dell Publishing, 1993), and E. D. Hirsch, ed., *What Your 3rd Grader Needs to Know* (New York: Doubleday, 1992).

9. Laura Ingalls Wilder, *Little House in the Big Woods* (New York: HarperCollins Publishers 1932), 178-79. This is the first of a wonderful series of fictionalized auto-biographies based on the life of a loving and adventurous pioneer family in the late nineteenth century.

10. A thorough and more leisurely exploration of the various meanings of fractions is presented in *Seeing Fractions: A Unit for the Upper Elementary Grades*, available from the California Department of Education, Sales, P.O. Box 271, Sacramento, CA, 95812.

    Both this and the Marilyn Burns' replacement unit mentioned in footnote 7 above are designed for classroom teachers but may interest parents, especially those who are home-schooling.

Chapter 11

1. Jack Price, "President's Report," *Journal for Research in Mathematics Education*, 27:5 (November 1996): 606.

2. See Appendix A for how to get this document.

3. Ohanian, *Garbage Pizza, Patchwork Quilts, and Math Magic*, 218, quoting Marilyn Burns.

4. The umpire and the catcher.

5. A lion and a lioness (see Ohanian, *Garbage Pizza*, 229).

6. The book *Family Math* (Jean Kerr Stenmark, Virginia Thompson, and Ruth Cassey, Berkeley, CA, 94720, Lawrence Hall of Science, University of California, 1986, page 48) gives directions for playing 7-Up and similar, harder games.

7. Nancy Casey and Michael R. Fellows, "Implementing the Standards: Let's Focus on the First Four," in *Discrete Mathematics in the Schools: How Can We Have an Impact? DIMAC/AMS Proceedings Series*, 1996.

8. Ohanian, *Garbage Pizza*, 230-31.

9. John Mason, keynote speech, annual NCTM meeting in Salt Lake City, April 1990. John Mason is an educator at Open University, Milton Keynes, England.

Chapter 12

1. President Clinton, quoted in *American Educator* (Spring 1996): 11-12.

2. Walter M. Haney, George F. Madaus, and Robert Lyons, *The Fractured Marketplace for Standardized Testing* (Boston: Kluwer Academic Publishers, 1993), 247.

3. Your state's Department of Education is located in your state's capital, but you can probably find its telephone number and address in your local telephone book, listed under "governments."

4. This quote is from page 2 of the October 1996 *Report No. 6* of the U.S. National Research Center announcing imminent publication of the book *Splintered Vision: An Investigation of U.S. Science and Mathematics Education* by Kluwer Academic Publishers. It will be available from U.S. TIMSS National Research Center, 455 Erickson Hall, Michigan State University, East Lansing, MI, 48824-1034. TIMSS is the acronym for Third International Mathematics and Science Study. Three other books about TIMSS will soon be published.

5. Bredkamp, ed., *Developmentally Appropriate Practices*.

6. Skeptics may also find comfort in the long footnote in Chapter 21 describing L. P. Benezet's experiments over 60 years ago with postponing all computational drill under grade six. Notice, however, that he provided most of the other NCTM activities in lower grades, and he did expect computational expertise by the time the youngsters graduated from eighth grade.

7. Casey and Fellows, "Implementing the Standards MACS/A 1996." This paper describes some topics from current mathematics suitable for the primary grades.

8. Fellows and Koblitz, "Combinatorially Based Cryptography," 14.

9. The classic book is *Math Their Way* by Mary Baratta-Lorton, published by Addison-Wesley Publishing Company in Reading, Mass., 1976 (reprinted in 1995).

10. T. Carpenter and J. Moser, "Developmentof Problem-Solving Skills," in *Addition and Subtraction: A Cognitive Perspective*, eds. T. Carpenter, J. Moser, and T. Romberg, Hillsdale, N.J.: (Lawrence Erlbaum Associates, Inc., 1982).

11. Penelope L. Peterson, Elizabeth Fennema, and Thomas Carpenter, "Using Knowledge of How Students Think About Mathematics," *Educational Leadership*, 46:4 (December 1988/January 1989): 42-46.

12. Susan L. Epstein, "Weighing Ideas," *Arithmetic Teacher* (April 1977): 293-97. This article describes teaching math concepts to second graders using a balance scale.

13. Mathematical Sciences Education Board, *Measuring Up: Prototypes for Mathematics Assessment* (Washington, D.C.: National Academy Press, 1993).

14. Deborah Loewenberg Ball, "The Mathematical Understandings That Prospective Teachers Bring to Teacher Education," *The Elementary School Journal*, 90:4 (March 1990), 449-466.

Chapter 13

1. Linda Darling-Hammond, *Beyond the Commission Reports: The Coming Crisis in Teaching*, R-3177-RC, Santa Monica, CA: Rand, July 1984.

2. Stevenson and Stigler, *The Learning Gap*, 164.

3. Ibid.

4. Invited speaker at the Joint Mathematics Meetings, Baltimore, Md., January, 1992.

5. Stevenson and Stigler, *The Learning Gap*, 171.

6. National Commission on Teaching & America's Future, *What Matters Most: Teaching for America's Future, Summary Report* (Woodbridge, Va., 1996), 15. The other countries are, in order, Belgium (with 80 percent), Japan, Italy, Australia, Finland, France, and Denmark (with slightly less than 60 percent).

7. Sykes, *Dumbing Down Our Kids*, 191.

8. David C. Berliner and Bruce J. Biddle, *The Manufactured Crisis: Myths, Fraud, and the Attack on America's Public Schools* (New York: Addison-Wesley Publishing Co., 1995), 264-65.

9. Ibid., 264.

10. "Public expenditures are positively related to state SAT and ACT performance," conclude Brian Powell and Lala Carr Steelman in "Bewitched, Bothered, and Bewildering: The Use and Misuse of State SAT and ACT Scores," *Harvard Educational Review*, 66:1 (Spring 1996), 27-59.

11. Samuel Freedman, *Small Victories* (New York: HarperCollins Publishers, 1990).

12. Seymour Fliegel with James Macguire, *Miracle in East Harlem: The Fight for Choice in Public Education* (New York: Random House, 1993), 26-27.

13. In *Crisis in Education: Stress and Burnout in the American Teacher* (San Francisco: Jossey-Bass Inc. Publishers, 1991). Barry A. Farber states that teachers were highly respected in the 1950s, but this eroded quickly in the '60s and deteriorated throughout the 1970s. Teachers see themselves as performing the most useful professional work and having the least prestige of twelve leading professions (1984 and 1989 polls, p. 206). A national public survey placed them third in usefulness, behind clergy and physicians, and fourth from the bottom in status, above local politicians, advertising practitioners, and realtors (1981 Gallup Poll, p. 205).

14. Stevenson and Stigler, *The Learning Gap*, 192.

15. Council of Chief State School Officers, *State Education Policies on K-12 Curriculum, Student Assessment, and Teacher Certification: 1995, Results of a 50-State Survey* (Washington, D.C.: Council of Chief State School Officers, 1995), 15.

16. National Commission on Teaching & America's Future, *What Matters Most: Teaching for America's Future, Summary Report* (Woodbridge, Va., 1996), 9.

17. Doctoral dissertation at Washington State University, Pulman, Wash., 1995. Dr Azin is now in the Mathematics Department of Eastern Washington University.

18. National Commission on Teaching & America's Future, *What Matters Most*.

19. Stevenson and Stigler, *The Learning Gap*, 150.

20. Ibid.

21. Ibid., 149.

22. Ibid., 150.

23. Farber, *Crisis in Education:* 196.

24. "Join the 'Par-aide' in Education: Volunteer parents in the classroom can enrich and extend the curriculum by sharing their career expertise, enthusiasm about avocations, and cultural knowledge," says Madeline Hunter in *Educational Leadership*, October 1989, pp. 36-41.

25. See "Computer Science Unplugged . . . Off-line Activities and Games for All Ages" in the bibliography. Fellow's articles include "Computer Science and Mathematics in the Elementary Schools," Mathematics and Education Reform, 1990-91, CBMS Issues in Mathematics Education 3 (1003): 143-163; "Combinatorially Based Cryptography for Children (and Adults)" in the Proceedings of the SE Conference on Graphs, Combinatorics, and Computing, 1993 (with Neil Koblitz); "Implementing the Standards: Let's Focus on the First Four," in Discrete Mathematics in the Schools: How Can We Have an Impact? DIMAC/AMS Proceedings Series, 1996 (with Nancy Casey); and "A Mathematics Popularization Event at CRYPTO '96," (with Neil Koblitz), 1996.

26. More discussion of public-school choice (with a stronger endorsement) appears on pages 176 and 177.

Chapter 14

1. Margaret Mead, *Blackberry Winter: My Earlier Years* (New York: William Morrow & Co., 1972), 46-47.

2. James W. Stigler and Harold W. Stevenson. "How Asian Teachers Polish Each Lesson to Perfection," *American Educator* (Spring 1991): 12-20 (especially pages 14-15).

3. Third International Mathematics and Science Study (TIMSS) as reported in *Splintered Vision: An Investigation of U.S. Science and Mathematics Education* (Boston: Kluwer Academic Publishers, 1996).

4. C. Silva and R. Moses, "The Algebra Project," *Journal of Negro Education*, 53:3 (1990): 378. This entire issue is about blacks in mathematics.

5. Jetter, "Mississippi Learning," sec. 6, 28-72.

6. C. Silva and R. Moses, "The Algebra Project," *Journal of Negro Education*, 53:3 (1990): 375-91.

7. *The Understanding Curriculum: Assessing U.S. Mathematics from an International Perspective* was the name of the report of the Second International Study of Mathematics, reflecting the authors' conclusions about what is *not* being taught in this country. The more recent Third International Study's report is *Splintered Vision*, which reflects the lack of focus in U.S. math curriculum (see Appendix A for references).

8. Sold by Cuisenaire; 800-237-0338. See Appendix D.

9. Sold by NCTM; 800-235-7566. See Appendix B.

10. AWM, 4114 Computer and Space Sciences Bldg., University of Maryland, College Park, MD, 20742-2461; 301-405-7892, AWM@math.umd.edu.

11. Available from NAACP, c/o Crisis Publishing Company, Inc., 4805 Mt. Hope Drive, Baltimore, MD, 21215; 800-781-5058 or 410-486-9169.

12. Available from 400 Washington Avenue, Montgomery, AL, 36104; Fax: 205-264-3121.

13. $57 annually, $28.50 for students; 800-235-7566.

14. Educational Testing Service, *America's Smallest School*, 36.

15. Sheila Tobias, *Math Anxiety* (New York: Norton, 1978), 1993.

16. Ibid.

17. Claudia Zaslavsky, *Fear of Math* (New Brunswick, N.J.: Rutgers University Press, 1994).

18. Marilyn Frankenstein, *Relearning Mathematics* (London: Free Association Books, 1989).

19. There was widespread publicity on November 21, 1996, the day after the results of the TIMSS study were released: for example, *The Star-Ledger*, p. 1. See Appendix A for references to the complete TIMSS study.

20. White. The bear must be a polar bear starting at the North Pole for the geometry to be correct.

21. Consider Roman numerals. Take the top half of XIII. This "problem" is a play on words. It illustrates the importance of understanding the language in which problems are posed, as emphasized in Chapter 8.

22. ACORN, *Secret Apartheid: A Report on Racial Discrimination Against Black and Latino Parents and Children in the New York City Public Schools* (Brooklyn, New York: ACORN Schools Office, 1996).

Chapter 15

1. Lewis H. Lapham, "Achievement Test," *Harpers Magazine* (July 1991): 10.

2. Ibid., 10, 13.

3. Ibid., 10-13.

4. Mathematical Sciences Education Board, *Everybody Counts*, 10.

5. John Chubb and Terry Moe, *Politics, Markets, and American Schools* (Washington, D.C.: Brookings, 1990).

6. Council of Chief State School Officers, *State Education Policies on K-12 Curriculum.*

7. Fliegel, *Miracle in East Harlem.*

8. Myron Lieberman, *Public School Choice: Current Issues/Future Prospects* (Lancaster, PA: Technomic Publishing Company, 1990).

9. Fliegel, *Miracle in East Harlem*, 90.

10. Adam Gamoran, "Do Magnet Schools Boost Achievement?" *Educational Leadership*, 54:2 (October 1996): 42.

11. National Coalition of Advocates for Students, "Student Admission and Placement," *The Good Common School: Making the Vision Work for All Children*, (Boston, 1991), 66.

12. *What Matters Most: Teaching for America's Future, Summary Report* available from the National Commission on Teaching & America's Future, P.O. Box 5239, Woodbridge, VA., 22194-5239 or Teachers College, Columbia University, Box 117, 525 West 120th Street, New York, NY, 10027.

13. ACORN, *Secret Apartheid.*

14. Joan Davis Ratteray, as quoted in *The Sunday Star-Ledger*, March 10, 1996, sec. 10, p. 11. The Institute for Independent Education (1313 N. Capitol St., NE, Washington, DC, 20002; 202-745-0500) conducts policy studies, especially about inner-city schools; engages in teacher development, especially in math; and has over 30 publications.

15. The movement to install computers almost constitutes a fourth approach to educational reform. Computers can indeed be helpful, but the extent to which computer salespeople are diverting money that is badly needed for helping teachers distresses me. A good teacher without a computer can teach excellent mathematics, but a good computer without a well-prepared, well-supported teacher is missing something vital. The three approaches discussed in more detail are more complex and less obviously connected to the profit motive.

Lisa Delpit in *Other People's Children: Cultural Conflict in the Classroom* (New York: The New Press, 1995, 95) speculates that white children both enjoy and learn from computers better than nonwhite children because their culture is more word-centered. She injects equity issues into the discussion of computers in education.

Computer Scientist Michael R. Fellows in "Computer Science and Mathematics in the Elementary Schools" (*Mathematics and Education Reform*, 1990-91, Conference Board of the Mathematical Sciences, Washington, D.C., 1993, 143-63) says, "It is a serious (and common) mistake to make a fetish of machines. Computer Science is not about machines in the same way that astronomy is not *about telescopes*. There is an essential unity of mathematics and computer science." His writing articulates well my own skepticism about the current rush to spend enormous amounts of tax money on machines, while neglecting people.

16. This would be a good time to reread about correlation and causation on pages 57-9. Given that high self-esteem and high math achievement often occur together, does this mean that raising math achievement causes better self-esteem or that raising self-esteem causes higher math achievement, or neither? It's not clear, but I vote for the first.

17. Kenschaft, "Successful Black Mathematicians of New Jersey," 4.

18. Council of Chief State School Officers, *State Education Policies on K-12 Curriculum*.

19. Sykes, *Dumbing Down Our Kids*.

20. Haney, Madaus, and Lyons, *The Fractured Marketplace*, xii.

21. Ibid., 208-9.

22. Ibid., 190.

23. Ibid., 189-294.

24. The state plans to further require a statewide fourth grade test beginning in 1998, at which time the state will withdraw its requirement for annual testing in grades 2-10.

25. Jonathan Kozol, *Savage Inequalities* (New York : HarperCollins, 1991), 143.

26. Ibid., 144.

27. Ibid., 144.

28. Ibid., 143-44.

29. See pages 225-28 for a discussion of how *much* time and money.

30. Stevenson and Stigler, *The Learning Gap*, 140.

31. Ibid., 141.

32. Ibid., 139.

33. Ibid., 141.

34. Ibid.

35. Jay Mathews, "Pst, Kid, Wanna Buy a . . . Used Math Book?" *Newsweek*, March 1, 1993. Education Section.

36. Saxon Publishers, Inc., 1320 West Lindsey, Norman, OK, 73069; 800-284-7019; 405-329-7071. There have been numerous popular articles about Saxon and his work. I received copies from this address, along with sample texts at varying levels. Although I was expecting Drill and Kill, I was pleasantly surprised. To me, it seems the Saxon phenomenon is one more piece of evidence that the NCTM principles are sound; Saxon has figured out one way to partially circumvent inadequate teacher preparation. Well-prepared teachers with inspiring texts could accomplish even more.

37. This passage was written before Saxon's fatal heart attack on October 17, 1996, but remains valid.

Chapter 16

1. Susan Epstein, "Learning Expertise from the Opposition—the Role of the Trainer in a Competitive Environment," *Proceedings of the Ninth Canadian Conference on Artificial Intelligence* (Vancouver: Morgan Kaufman, 1992), 236.

2. E. D. Hirsch, ed., *What Your 2nd Grader Needs to Know*, 252.

3. Fresh Fruit Records, 369 Montezuma #209, Santa Fe, NM, 87501; 1-800-is-fruit.

4. J. Arthur Jones, "Blacks in Science: A Growing National Crisis," Proceedings of the Eleventh Annual Meeting of NAM, San Antonio, Texas, January 3-4, 1980, 22-23.

5. Organizations are springing up locally and nationally of parents and other adults who insist that all children should benefit from the uplifting educational practices that have too often been offered only to the "gifted." Some of these organizations cater to particular groups and others are broad-based. See Appendix A for their policy statements and Appendix B for brief descriptions of the groups themselves.

6. Bredekamp, ed., *Developmentally Appropriate Practices*, 66. Many research references are provided for the NAEYC assertion, supporting the claims of this chapter. I believe that excessive drill deadens the abilities of older children and adults (almost?) as much as those of young children. These references, however, refer primarily to children in the primary grades.

Chapter 17

1. Edward Lee Thorndike, *The Psychology of Arithmetic* (1922) as quoted in *Psychology and the Science of Education: Selected Writings of Edward L. Thorndike* (New York: Teachers College Press, Columbia University, 1962), 84.

2. Ibid., 84.

3. Edward L. Thorndike and Arthur I. Cates, *Elementary Principles of Education* (New York: Macmillan, 1931), 257.

4. Ibid., 258.

5. In each of his books *What Your nth Grader Needs to Know* (n = 1, 2, 3, 4), E. D. Hirsch writes "The three cardinal principles of elementary mathematics education are 1) practice, 2) practice, 3) practice. . . . " See also page 34 in Chapter 3.

6. National Commission on Teaching and America's Future, *What Matters Most, Summary Report*, 10.

7. As E. D. Hirsch and too many texts define it.

8. Ohanian, *Garbage Pizza*, 85-86.

9. A devastating expose of how financial pressures to make quick, cheap standardized tests without sufficient "validity" (i.e., "the accuracy of inferences drawn from the results of the test" [p. 190]) are polluting many types of high-stake decisions in our society appears in *The Fractured Marketplace for Standardized Testing*, Walter M. Haney, George F. Madaus, and Robert Lyons (Boston: Kluwer Academic Publishers, 1993), especially the chapter entitled "Test Quality."

Chapter 18

1. Haney, Madaus, and Lyons, *The Fractured Marketplace*, 286, 294.

2. Ibid., and Daniel Koretz, Brian Stecher, Stephen Klein, and Daniel McCaffery, "The Vermont Portfolio Assessment Program, Findings and Implications," *Educational Measurement: Issues and Practice*, 13:3 (Fall 1994): 5-16.

3. Mathews, *Escalante*, 48.

4. Sykes, *Dumbing Down Our Kids*, 150-51.

5. Koretz, Stecher, Klein, and McCaffery, "The Vermont Portfolio Assessment Program," 5-16.

6. Daniel Koretz, Brian Stecher, Stephen Klein, and Edward Diebert, "Can Portfolios Access Student Performance and Influence Instruction? The 1991-92 Vermont Experience" (Los Angeles, Calif., CRESST, 1993). Also, the same co-authors in "Interim Report: The Reliability of Vermont Portfolio Scores in the 1992-93 School Year" (Santa Monica, Calif., RAND, 1993).

7. *Improving America's Schools: A Newsletter in School Reform* (Spring 1996).

8. Chantal Shanroth, "A Comparison of University Entrance Exams in the United States and Europe," *Focus*, 13(3)(1993): 1, 11-14.

9. National Council of Teachers of Mathematics, *Assessment Standards for School Mathematics*, 1995, 60.

10. Haney, Madaus, and Lyons, *The Fractured Marketplace*, 293.

11. A fascinating international history of testing and analysis of various testing forms is found in the chapter "Curriculum Evaluation and Assessment" by George F. Madaus of Boston College and Thomas Kellaghan of the Educational Research Centre, Dublin, Ireland, in the *Handbook of Research on Curriculum*, ed., Philip W. Jackson (New York: Macmillan, 1992).

12. Not applicable in revised edition of *Math Power*.

13. Not applicable in revised edition of *Math Power*.

14. Commonly available samples of such charts on which you can record young children's progress accompany *Mathematics Their Way* by Mary Baratta-Lorton.

15. George H. Wood, *Schools That Work: America's Most Innovative Public Education Programs* (New York: Dutton, 1992), 32.

16. Alan Gartner and Dorothy Kerzner Lipsky, "Beyond Special Education: Toward a Quality System for All Students," *Harvard Educational Review*, 57:4 (November 1987): 367-95.

17. I strongly agree with the position of the NAEYC in their repeated insistence that retention and placement in special or remedial classes be based primarily on information obtained from observations by teachers and parents, not on the basis of a single test score. The association's views appear in *Developmentally Appropriate Practices in Early Childhood Programs Serving Children from Birth Through Age 8*; 57, 76. This is just as valid for older children and adults except that parental observation is gradually replaced by self-observation.

18. Patricia C. Kenschaft, "Charlotte Agnas Scott, 1858-1931," *College Mathematics Journal*, 18:2 (March 1987): 98-110. Also, Patricia C. Kenschaft, "Charlotte Agnas Scott (1858-1931)," *Women in Mathematics: A Bibliographic Source Book*, eds. Louise S. Grinstein and Paul Campbell (Westport, Conn.: Greenwood Press, Inc., 1987), 193-203; reprinted in *A Century of Mathematics in America, Part III* (Providence: American Mathematical Society, 1988).

Scott was the first Mathematical Department chair of Bryn Mawr College, from 1885 to 1925, and published widely in analytical geometry. Her text in the field was used for decades. She was on the first board of the American Mathematical Society and served as its vice president. Already "decidedly hard of hearing" but nevertheless an excellent teacher according to her 1885 recommendations, she became totally deaf by early in the twentieth century, but was an indefatigable lip reader.

19. "Academician Lev Semenovich Pontryagin (Obituary)," *Russian Math Surveys*, 44:1 (1989): 1-2; and other previous articles.

Pontryagin (1908-1988) was president of the USSR commission on school mathematical education. His 1938 book *Topological Groups* (translated) was a standard text during my own graduate work in pure mathematics. Later, his "Pontryagin Optimization Principle" revolutionized applied mathematics, and he led the department of optimal control at Moscow State University.

20. Given less than three years to live when he contracted Lou Gehrig's disease in 1962, Stephen Hawking (b. 1942) is still producing. He has written many scholarly articles and several books, the most accessible of which are *A Brief History of Time* (1988) and *Black Holes and Baby Universes* (1993), both published by Bantam Books.

Both contain autobiographical information. Popular pieces about him have appeared in *People* (September 11, 1989) and *Time* (February 8, 1988; August 31, 1992; and September 27, 1993). A more scholarly summary of his life and achievements appears in *Contemporary Authors*, volume 48, Pamela Dear, ed., Gale Research, Inc., Detroit, MI, 1995. 224-228.

21. Fliegel, *Miracle in East Harlem*, 118.

22. Leonhardt, *Parents Who Love Reading, Kids Who Don't*.

23. National Coalition of Advocates for Students, *The Good Common School*, 84.

24. Collins and Tamarkin, *Marva Collins' Way*.

25. Ibid., 188-89.

26. Deborah Meier, "Success in East Harlem: How One Group of Teachers Built a School That Works," *American Educator*, 11:3 (Fall 1987): 34-39.

27. Fliegel, *Miracle in East Harlem*, 161.

28. Other successful examples are described on pages 74-80 of *The Good Common School: Making the Vision Work for All Children*, in the section entitled "Student Admission and Placement," National Coalition of Advocates for Students, Boston, Mass., 1991.

29. National Commission on Teaching & America's Future, *What Matters Most, Summary Report* 25.

30. National Council of Teachers of Mathematics, *Assessment Standards for School Mathematics*.

31. Copies of Goals 2000 and other national documents are easy to locate on the U.S. Department of Education's web site at http://www.ed.gov.

32. College Board, *Counselor's Handbook for the SAT Program*, 1996-97 (College Board, New York, 1994), Table 9, 52.

33. Sykes, *Dumbing Down Our Kids*, 148-49.

34. College Board, *Counselor's Handbook*, Table 19, 52.

35. Internal memo from ETS.

36. Other implications of the change, the reason for it, and a brief history of the SATs are discussed in "Have Changes in the SAT Affected Women's Mathematics Performance?" by Nancy Burton of the Educational Testing Service in *Educational Measurement: Issues and Practice*, 15:4 (Winter 1996), 5-9.

37. "No letter or numerical grades are given during the primary years," insists the National Association of the Education of Young Children in its guidelines document *Developmentally Appropriate Practices in Early Childhood Programs Serving Children from Birth Through Age 8*, 75.

Chapter 19

1. John Holt, *Why Children Fail* (Reading, Mass.: Addison-Wesley, 1964, 1982), 70 (first quote), 5 (second quote).

2. Haney Madaus, and Lyons, *The Fractured Marketplace*, 197.

3. Jesus, *Matthew* 7:1, King James Bible.

4. John Jacob Cannell, "Nationally Normed Elementary Achievement Testing in America's Public Schools: How All 50 States Are Above the National Average," *Educational Measurement: Issues and Practice*, 7:2 (Summer 1988).

5. Mathews, *Escalante*, 227.

6. Ibid., 217.

7. Ibid., 264.

8. Amy Stuart Wells and Irene Serna, "The Politics of Culture," *Harvard Educational Review* (1996), 96.

9. Ibid., 106.

10. Sykes, *Dumbing Down Our Kids*, 241-260.

11. Ibid., 144.

12. Ibid., 143-51.

13. Haney, Madaus, and Lyons, *The Fractured Marketplace*, 36.

14. Ibid., 189-246.

15. Marilyn Burns, speaking at the NCTM Regional Conference, Parsippany, N.J., November 1994.

16. Mary L. Smith, "Put to the Test: The Effects of External Testing on Teachers," *Educational Researcher*, 20:5 (1991): 8-11; Mary L. Smith, "Meanings of Test Preparation," *American Educational Research Journal*, 28:3 (1991): 521-41.

17. Thomas Toch with Betsy Wagner, "Schools for Scandal," *U.S. News and World Report*, 27 April 1992, 70.

18. W. Haney, G. Madaus, and R. Lyons, *The Fractured Marketplace*.

19. Ibid., 6.

20. Stevenson and Stigler, *The Learning Gap*, 205-6. For comparisons with other countries, some of which spend as much as 80 percent of their school budgets on classroom teaching, see p. 148.

21. J. J. Cannell, *The 'Lake Wobegon' Report: How Public Educators Cheat on Standardized Achievement Tests* (Albuquerque: Friends for Education, 1989), quoted in *The Fractured Marketplace* by Haney, Madaus, and Lyons, 193.

22. M. L. Smith, "Meanings of Test Preparation," *American Educational Research Journal*, 28:3, 541-42.

23. Ibid.

24. Thomas Toch with Betsy Wagner, "Schools for Scandal," *U.S. News and World Report*, 27 April 1992, 69; Gary Putka, "A Cheating Epidemic at a Top High School Teaches Sad Lessons," *The Wall Street Journal*, 29 June 1992, 1, 4, 5.

25. Haney, Madaus, and Lyons, "Test Quality," *The Fractured Marketplace*, 189-245.

26. Mathews, *Escalante*, 143-79.

27. Phyllis Rosser, *The SAT Gender Gap: Identifying the Causes* (Washington, D.C.: Center for Women Policy Studies), $15.

28. Cathy Kessel and Marcia C. Linn, "Grades or Scores: Predicting Future College Mathematics Performance," *Educational Measurement: Issues and Practices*, 15:4, (Winter 1966): 11. Another source by the same authors is in *Research in Collegiate Mathematics Education* (vol. 2, 1996), edited by J. Kaput, A. Schoenfeld, and E. Dubinsky, and published by the American Mathematical Society in Providence, R.I.

29. C. S. Dweck, W. Davidson, S. Nelson, and B. Emma, "Sex Differences in Learned Helplessness. Part 2: The Contingencies of Evaluative Feedback in the Classroom," *Developmental Psychology*, 14:3 (1978): 268-76; Elizabeth Fennema and Penelope Peterson, "Autonomous Learning Behavior: A Possible Explanation of Gender-Related Differences in Mathematics," *Gender-Related Differences in Classroom Interaction*, eds. L. C. Wilkinson and C. B. Marret (New York: Academic Press, 1985), 17-36.

    For further references and a list of 55 ways that our culture discourages women and girls in mathematics, see *Winning Women into Mathematics*, ed. Patricia Clark Kenschaft, Mathematical Association of America, Washington D.C., 1991.

30. An erudite exposition of these issues is available in the *Harvard Educational Review* (64:1, Spring 1994, 76-95) in an article entitled "A Technological and Historical Consideration of Equity Issues Associated with Proposals to Change the Nation's Testing Policy," and written by George Madaus.

31. George F. Madaus, "The Effects of Important Tests on Students: Implications for a National Examination System," *Phi Delta Kappan* (November 1991): 227.

Chapter 20

1. People are searching Helen Keller's thousands of articles for the precise citation, but this quote is "definitely Helen's," according to Allison Bergmann, Librarian, American Foundation for the Blind, 1-800-232-5463.

2. Ann Kahn, "Math Matters," *PTA 89, A Special Redbook Supplement*, 13-14.

3. Beverly T, Watkins, "Many Campuses Now Challenging Minority Students to Excel in Math and Science," *Chronicle of Higher Education* (June 14, 1989): A13, A16.

4. Uri Treisman and Rosa Asera, "Teaching Mathematics to a Changing Population: The Professional Development Program at the University of California, Berkeley. Part 1: A study of the Mathematics Performance of Black Students at the University of California, Berkeley, Part 2: The Math Workshop: A Description," *Mathematics and Education Reform*, Proceedings of the July 6-8, 1988, Workshop, eds., Naomi Fisher, Harvey Keynes, Philip Wagreich, American Mathematical Society, Providence, R.I., 1990, 11-62.

5. Marvin Cetron and Margaret Gayle, *Educational Renaissance, Our Schools at the Turn of the Twenty-First Century* (New York: St. Martin's Press, 1991), 120.

6. Owen Tregaskis, "Parent and Mathematical Games," *Arithmetic Teacher* (March 1991): 14-16. One game, for example, is "Make Six!" Take the aces (as ones), twos, threes, fours, and fives from two decks of standard playing cards, shuffle, and give half to each player. Players take turns putting the cards face-up in two piles on the table. The first player to recognize that the two cards showing add to six calls out "Make Six!" and gains all the cards on the table. The winner is the first to gain all the cards. Parents can decide when to play "Make Seven!" and "Make Eight!"

7. Mary T. Hoes and Joy Faini Saab, "Where All Are Winners: A Mathematics Olympics for Parents, Students, and Teachers," *Teaching Children Mathematics*, 3:3 (November 1996): 118-21. Numerous references to articles with related ideas are included.

8. Frederick L. Silverman, Ken Winograd, and Donna Strohauer, "Student-Generated Story Problems," *Arithmetic Teacher*, 39:8 (April 1992): 6-12.

9. Dr. Joyce O'Halloran (Department of Mathematical Sciences, Portland State University, Portland, OR, 97207-0751) has written a manual for people who want to implement this idea in their own community. It is available for $5.

10. Education leader Madeline Hunter urges teachers to invite parents to their classrooms to share their careers, hobbies, and cultures. See, for example, her article, "Join the 'Par-aide' in Education: Volunteer Parents in the Classroom Can Enrich and Extend the Curriculum by Sharing Their Career Expertise, Enthusiasm About Avocations, and Cultural Knowledge," *Educational Leadership*, October 1989, 36-41.

11. *Family Math*, Lawrence Hall of Science, University of California, Berkeley, CA, 94720; 510-642-5132; http://www.lhs.berkeley.edu/.

12. A similar program is "Math Alive," established by the Institute for Independent Education for teachers. The Institute sees math as the key subject in empowering minority children to attain satisfying places in society. You can contact the organization at 1313 N. Capitol St., NE, Washington DC, 20002; 202-745-0500.

13. Jetter, "Mississippi Learning," sec. 6, 28ff.

14. C. Silva and R. Moses, "The Algebra Project," *Journal of Negro Education*, 53:3 (1990): 375-91. This entire issue is about blacks in mathematics.

15. Ibid., 376-77.

16. Ibid., 390.

17. See the work of Michael Fellows, listed in Note 25, Chapter 13.

18. Jean K. Stenmark, Virginia Thompson, and Ruth Cassey, *Family Math* (Berkeley: University of California at Berkeley [Lawrence Hall of Science], 1986) in English and Spanish.

19. Now, every state's compulsory education legislation explicitly allows home-schooling, and only Michigan requires some involvement of certified teachers. All states do require that families file with a state or local agency, explaining their plan for home-schooling. The only federal guideline has been a limited Supreme Court decision (Wisconsin vs. Yoder, 1972), involving the Amish. About a half million school-aged children are home-schooled each year. This is about 1% of the school-aged population and about 10% of the privately educated population. Since many children are home-schooled only temporarily, significantly more than 1% of the population is home-schooled at some time during their K-12 education (from *ERIC Digest*, number 95, by Patricia Lines).

   An accessible article about home-schooling appeared in the *New York Times Magazine* on February 2, 1997, on pages 30-37.

Chapter 21

1. Daniel McGaurty, *Break These Chains* (00:00, 1996) [cover quote from an interview with Williams].

2. The National Commission on Teaching and America's Future, *What Matters Most: Summary Report 5* (first two quotes), 7 (third quote), 8 (fourth quote). (See Appendix A.)

3. Burrill's response, in a media release the day after the results of the Third International Mathematics and Science Study (see Appendix A) were made public. (For example, *The Star-Ledger*, November 21, 1996.)

4. If you become deeply involved in local structural change, you will want to become acquainted with the innovative work of James Comer, John Goodlad, Henry Levin, and Theodore Sizer in renovating schools and districts. Their (very significant) contributions are beyond the scope of this book.

5. Mathematical Sciences Education Board, *Everybody Counts*.

6. Stevenson and Stigler, *The Learning Gap*, ch. 6.

7. I find great hope and joy in the report *What Matters Most: Teaching for America's Future*, which was released after this book was written but before it was published. See Appendix A; reading it would be a fine follow-up to reading *Math Power*.

8. National Commission on Teaching & America's Future, *What Matters Most*.

9. *School Power* (New York: Free Press, a division of Macmillan, 1980) give a dramatic, moving, and at times hair-raising account of Comer's first (and successful) efforts toward school reform. It provides insight into the daily troubles of real teachers in our country.

10. James P Comer, "Educating Poor Minority Children," *Scientific American*, 259:5 (November 1988): 42-48.

11. National Alliance for State Mathematics and Science Coalitions, 1527 18th Street, NW, Washington, DC, 20036; 202-387-3600; Fax: 202-387-4025.

12. Coordinated by the Mathematical Sciences Education Board, 2101 Constitution Avenue, NW, HA 450, Washington, DC, 20418; 202-334-3294; Fax: 202-334-1453.

13. Many translations and editions of Alexis de Tocqueville's two-volume *Democracy in America*, based on his trip to this country, have appeared since the 1840s. One of the most recent is from Anchor Books, Garden City, N.Y., 1969.

14. Patricia C. Kenschaft, "Mathematicians and Elementary School Teachers Need Each Other," *Proceedings of the 44th International Meeting of ICSIMT* (International Commission for the Study and Improvement of Mathematics Teaching) 1992, University of Chicago, Dept. of MCS, Box 4348, Chicago, IL, 60680, 55-64.

15. However, if your child is young, pre-service is of personal interest to you, too. Half the U.S. K-12 teachers in 2006 will join the profession in the coming decade, according to *What Matters Most: Teaching for America's Future* by the National Commission on Teaching & America's Future page 7 of the *Summary*.

16. This position is tame compared to that of Manchester, N.H., school superintendent L. P. Benezet, who suspected over 60 years ago that children would know arithmetic better at the end of eighth grade if they weren't taught any "tables" or algorithms until sixth grade. He verified his view by comparing children taught the traditional curriculum with those in classes where the three R's were reading, reasoning, and reciting. ". . .The effect of the early introduction of arithmetic had been to dull and almost chloroform the child's reasoning faculties." He believed that extensive language development (see Chapter 8) should precede learning arithmetic skills, but he did encourage teachers to have children play games involving numbers and shapes, to measure everything, to reason and explain their reasoning orally, and to explore whatever math was actually needed in their daily lives, including their reading assignments. His dramatic results are reported in three articles in *The Journal of the National Education Association*, 24:8 (November 1935): 241-44; 24:9 (December 1935): 301-3; and 25:1 (January 1936): 7-8.

17. Harriet Tyson, *Who Will Teach the Children: Progress & Resistance in Teacher Education* (San Francisco: Jossey-Bass Publishers, 1994), 137.

18. Rita Kramer, *Ed School Follies: The Miseducation of Teachers* (New York: The Free Press, a division of Macmillan, 1991).

19. Tyson, *Who Will Teach the Children?*, 143.

20. E. D. Hirsch, *What Your 1st Grader Needs to Know* (New York: Dell Publishing, 1993), 158.

21. *Indicators of Mathematics and Science Education* (Washington, D.C.: National Science Foundation, 1995), 39, 136.

22. Christine Keitel, "Mathematics Education and Educational Research in the USA and USSR: Two Comparisons Compared," *Journal of Curriculum Studies*, 14:2 (1982): 111.

23. McKnight et. al., *The Underachieving Curriculum: Assessing U.S. School Mathematics from an International Perspective, A National Report on the Second International Mathematics Study* (Champaign, Ill.: Stipes Publishing Company, 1987), 26.

24. Ibid.

25. W. Servais and T. Varga, *Teaching School Mathematics* (Baltimore, Md.: Penquin Books-Unesco, 1971), 236.

26. The National Network of Partnership-2000 Schools (see p. 269) helps families, schools, and communities build partnerships. Director Joyce Epstein has written two balanced pieces about what partnerships can and cannot do, which *Math Power* readers may want to consult: "School-Family-Community Partnerships: Caring for the Children We Share," *Phi Delta Kappan* (May 1995): 701-12; and "Perspectives and Previews on Research and Policy for School, Family, and Community Partnerships" in *Family School Links: How Do They Affect Educational Outcomes?*, eds. Allan Booth and Judith Dunn (Mahwah, N.J.: Lawrence Erlbaum Associates, 1996).

27. See the comments on p. 210 about the leadership of North Carolina in eliminating such testing.

Chapter 22

1. Kenschaft, "Black Women in Mathematics in the United States" (October 1981): 592-604; reprinted with photographs and postscript in the *Journal of African Civilization*, 4:1 (April 1992): 63-83.

2. David Gale, "The Mathematical Millennium," *Notices of the American Mathematical Society,* Providence, R.I., 43:12 (December 1996): 1518-19.

3. Douglas Hofstadter, *Godel, Escher, Bach* (New York: Basic Books, 1979).

4. Special math materials catering to African-American students are proliferating. For example, the Bingwa Software Company published the Mathematical Heritage Series in 1994, which integrates elementary grade-level math into the lives of outstanding African Americans.

5. Leading educators made strong statements supporting this view in the 1996 report *What Matters Most: Teaching for America's Future*, available via a google search on the title.

6. Duane A. Cooper, "A Gateway to Opportunities," *Math Horizons* (April 1996): 20-24.

Index

employment opportunities in math. *See* careers

environmental issues, 6, 9–10, 123–124, 247, 265

Epstein, Susan, 73, 187–188

equations, 17, 33, 44–46, 161

equity issues, 5, 25, 149, 150, 173–174, 176–178, 181, 192–194, 211–215, 238. *See also* African American mathematicians

  defying predictions, 4, 15, 160, 189, 205, 213, 214–215, 218, 223–224, 243

  elitism, 125, 149, 173, 224

  financial disparity, 149, 224–225

  language, 82-83, 93

  racism, 4, 7, 64, 149, 173–174, 176–177, 181, 192–194, 213–215, 224, 230

  resources, 281–282

  sexism, 59, 99, 134, 166, 194, 230, 260

  suggestions, 72, 99, 103–104, 166, 237, 241–243, 258–259

  testing, 4, 205, 208–209, 223, 230–232, 256

  textbook problems, 191–192

  tracking, 214–215, 223–224, 232

  Zaslavsky books, 167

Escalante, Jaime, 15, 33, 223

Escalante: The Best Teacher in America (Mathews), 223, 229, 279

estimation, 56, 129

ETS. *See* Educational Testing Service (ETS)

Everybody Counts: A Report to the Nation on the Future of Mathematics Education, 3, 69, 237, 247, 266

Everyday Mathematics, 270

excellence in schools, 152, 154, 174–175, 248

exercise (physical), 24

exercises (math), 159–160. *See also* drill

expectations, 64–65, 175

Experiencing School Mathematics: Teaching Styles, Sex and Setting (Boaler), 279

EXXON Education Foundation, 33

facts, mathematical, 126, 129, 140–141, 200–202. *See also* arithmetic; games

Failing at Fairness: How America's Schools Cheat Girls (Sadker), 282

failure, 18–20, 30–31, 192

fairs, math, 238–239

FairTest: National Center for Fair and Open Testing, 282

families, 1, 5, 103

Family Math book and program, 239–240, 270, 275, 276

Family Pastimes (game publisher), 273

Family Things (publisher), 271

Fear of Math (Zaslavsky), 167, 281

feelings, 8, 15–20, 22, 25–26, 30–31, 34–35, 37, 39, 56–60, 98, 178, 192–196. *See also* joy

Fellows, Michael, 8, 12, 127, 136–137, 156

Fennema, Elizabeth, 139

Fermat, Pierre, 17

fewer and less, 87–88

fifth grade, 133, 141–143, 159–170

finances, 11, 148–149, 183, 203, 226–228

first grade, 50–51, 64, 89–90, 98–99, 137–139

flexibility, 28–29, 80–81, 93–94, 108–110, 257

Fliegel, Seymour, 176, 213

food and allergies, 57

forgetting, 31–32, 81

fourth grade, 141

fractions, 103, 120–124, 131–134, 140–141

  age learned, 26, 133, 139–143

  on the number line, 44, 132, 140

  reducing, 132–133

The Fractured Marketplace for Standardized Testing, 135, 180, 201, 205, 208, 226–229, 282

fragmented lives and learning math, 12, 265

Frankenstein, Marilyn, 167

free will, 25

frustration, 18–20, 31, 35, 37, 139, 195

FUN Books (catalog), 271

fun drill, 69–71, 77–79, 91, 184

fun with math, 3, 7–8, 14–15, 29, 35, 39, 47, 52, 63–66, 68–69, 191, 236, 258

funding of schools, 148–149, 207–209, 226–228

Women in Mathematics, 283
women mathematicians, books about, 283.
        See also sexism
Wonderful Ideas, Having of, 29, 36, 38
words, 84–94. *See also* definitions; *ganas*
  slipperiness of, 38, 83, 88
  undefined terms, 98–99
work or play?, 108–109
wrong answers, 50–52, 258. *See also* right
        answers
  admitting to being wrong, 18
  all right to be wrong, 47, 52, 81
  responding to, 37, 41–42, 52–57
  in textbooks, 268

Young Children Reinvent Arithmetic,
       Implications of Piaget's Theory
       (Kamii), 279

Zaslavsky, Claudia, 167, 281
zero, 44–45, 51